The First Outstanding 50 Years of "Università Politecnica delle Marche"

Sauro Longhi · Andrea Monteriù ·
Alessandro Freddi · Giulia Bettin ·
Silvio Cardinali · Maria Serena Chiucchi ·
Marco Gallegati
Editors

The First Outstanding 50 Years of "Università Politecnica delle Marche"

Research Achievements in Social Sciences and Humanities

 Springer

Editors
Sauro Longhi
Department of Information
Engineering (DII)
Marche Polytechnic University
Ancona, Italy

Andrea Monteriù
Department of Information
Engineering (DII)
Marche Polytechnic University
Ancona, Italy

Alessandro Freddi
Department of Information
Engineering (DII)
Marche Polytechnic University
Ancona, Italy

Giulia Bettin
Department of Economics and Social
Sciences (DISES)
Marche Polytechnic University
Ancona, Italy

Silvio Cardinali
Department of Management (DM)
Marche Polytechnic University
Ancona, Italy

Maria Serena Chiucchi
Department of Management (DM)
Marche Polytechnic University
Ancona, Italy

Marco Gallegati
Department of Economics and Social
Sciences (DISES)
Marche Polytechnic University
Ancona, Italy

ISBN 978-3-030-33881-7   ISBN 978-3-030-33879-4   (eBook)
https://doi.org/10.1007/978-3-030-33879-4

This Springer imprint is published by the registered company Springer Nature Switzerland AG
The registered company address is: Gewerbestrasse 11, 6330 Cham, Switzerland

Foreword

Università Politecnica delle Marche (UNIVPM in short) celebrates its 50th anniversary as an institution in 2019. It is a young, dynamic, and active University on multiple fronts, which in the first half-century of its life has been able to achieve important goals in many scientific areas. Hence, the idea of proposing a book is for celebrating the first 50 years, "UNIVPM50", and the natural choice is to divide the book into three thematic volumes: "Social Sciences and Humanities", "Physical Sciences and Engineering", and "Life Sciences".

The contents of these three volumes document the research activities of our University, both fundamental, applied, and clinical. "The ability to think together and to produce research, through differentiated thought and culture collaboration, is the main reason for the existence of an academic structure"—from "L'Università di Ancona, 1969/1989", a book celebrating the first 20 years of our University. The objective of "UNIVPM50" is therefore to present the most important research results achieved so far by our scientific community, with particular attention to the current frontiers of research and what future prospects might be. The contributions collected in the book summarize the main research efforts undertaken by the current faculty members and represent the outcome of a constant effort in the present, supported by past experiences, with the perspective to new emerging ideas in the three scientific areas in which UNIVPM plays a significant role.

The book, however, is not to be considered only as a collection of research works: the contributions also represent a sample of the knowledge which is daily taught in our classrooms and laboratories. Indeed, from the research, the University constantly draws inspiration for high-quality teaching and has the great responsibility of preparing young people for the world of professions. To train professionals, it is necessary that the link between academia and professional world is strong. The relationship with the territories is strengthened with actions to boost research and innovation, the so-called third mission, made up of technology transfer, patents, and entrepreneurship. These activities have led to the development of a contamination lab, followed by an incentive and promotion policy for university spin-offs that draw ideas from the best research results.

Founded 50 years ago as "University of the territory", UNIVPM must now confront itself in a perspective as broad and international as possible. Our research was, and still is, inspired by the territories, with the aim of increasingly open to the world. Internationalization is therefore a key to the present and future development of our University: international research, active student/teacher mobility policies, and the start of international degree courses, even as double degrees, are all measures aimed at projecting the University into an international dimension.

In conclusion, "UNIVPM50" presents the main goals achieved in the first 50 years of story of our University, and imagine a future of which UNIVPM will, hopefully, be an integral part.

Ancona, Italy   Sauro Longhi
                                        Rector of the Università Politecnica delle Marche

Preface

Università Politecnica delle Marche (UNIVPM in short) celebrates its 50th anniversary as an institution in 2019.

Historically, UNIVPM was first established in Ancona in 1969 as Faculty of Engineering, followed by the Faculty of Medicine the next year. In 1971, the university was granted the status of State University, with the name of "University of Ancona". The Faculties of Engineering and Medicine were later merged with the Faculty of Economics and Commerce in 1982. Further on, in 1988 and in 1991, respectively, the Faculty of Agriculture and the Faculty of Sciences were established. In 2003, the name of the university was changed to "Università Politecnica delle Marche".

Focusing on the present, the Faculty of Engineering is composed of four departments: "Construction, Civil Engineering and Architecture", "Materials, Environmental Sciences and Urban Planning", "Information Engineering" and "Industrial Engineering and Mathematical Science". The Faculty of Medicine and Surgery has four departments: "Biomedical Science and Public Health", "Experimental and Clinical Medicine", "Molecular and Clinical Sciences" and "Odontostomatologic and Specialized Clinical Sciences". The Faculty of Economics is composed of two departments: "Management" and "Economics and Social Sciences". Finally, the Faculties of Agriculture and Sciences count a department each, "Agriculture, Food and Environmental Sciences" and "Life and Environmental Science", respectively.

UNIVPM placed fifth in the 2018/2019 Censis Italian University Ranking among the medium-sized public universities. Being a "young university", this must be considered as a good result. On the other hand, there is still a lot of work to do in order to position UNIVPM at higher levels of research, but thanks to our growing faculty's expertise remarkable progress is being made each year.

This book is dedicated to the celebration of 50 years of UNIVPM and has been motivated by the desire to present the most representative research results of our scientific community achieved so far, with particular attention to the current frontiers of research and to what the future perspectives could be, in order to build a better society providing a dynamic road map for the future. The contributions collected in the book summarize the main research efforts undertaken by current faculty members, from a vast range of researches. They represent the outcome of a constant effort in the present, supported by past experiences, with the perspective to new emerging ideas in the three major cores where UNIVPM plays a significant role, namely Life Sciences, Physical Sciences and Engineering and Social Sciences and Humanities.

In the present volume, the scientists working at the UNIVPM in the Social Sciences and Humanities branch describe the major results obtained in their research work and the expected future progresses.

The development and the success of the Faculty of Economics of UNIVPM are mainly due to Giorgio Fuà, who personally guided all the activities starting from the recruitment of the teaching staff. Over the years, the far-sighted recruitment policy adopted by Fuà brought in Ancona the best scholars in various disciplines. Giorgio Fuà was so relevant for the faculty that it was named after him in 2002.

The aspects that characterized the faculty and allowed it to stand out at the national level are closely linked both to Fuà's activities as an applied economist and to his research interests. Fuà's contribution was characterized by the will to interpret the process of Italian economic development and its territorial aspects from a historical perspective, the attention to the concrete problems of his time, the attempt to combine the political economy and the social reforms. These distinguished features were reflected in the contamination between scholars from different disciplines and in the creation of a strong link with the territory that led to the foundation in 1966 of the Institute for Economic and Social Studies Adriano Olivetti, later renamed ISTAO. Fuà was its founder, manager, factotum, president and financial promoter.

The strong openness to interdisciplinary skills, in contrast to the currently predominant thematic specialization, and the contribution to the economic and civil development of the society continue to be relevant aspects, and this is highlighted by the strategic projects funded by UNIVPM and the attention to activities related to relationships with the territory, the so-called Third Mission.

In line with this interdisciplinary approach, the present volume starts with a contribution by Chelli, Gregori, Marasca and Papi that highlights the main trends in teaching and research activities carried out at the Faculty of Economics from 1959 onwards in four main areas: Business Administration, Economics and Business Management, Economic Statistics and Economics.

Gallegati, Palestrini and Russo discuss the origin of the agent-based model's school in Ancona. Heterogeneity of economic individuals and their interactions mediated by networks were introduced in traditional economic modelling in order to analyse complex systems.

The contribution by Chiapparino, Ciuffetti and Giulianelli describes the main research topics in economic history at the Faculty of Economics "Giorgio Fuà". The original and innovative nature of these research activities contributed to the prominent position gained by University of Ancona in the recent Italian historiographic debate.

Along the same line, Ascoli, Carboni and Vicarelli explain how sociological studies developed in Ancona. Starting from the economic development, the range of topics widened rapidly by including public policies and the welfare state, healthcare professionalism and civil and social mobilization.

The experience of the Money and Finance Research Group is depicted by Alessandrini, Papi and Zazzaro along three main strands of research: the international and European monetary systems, the relationship between the banking structure and local development and the potential discriminatory effects of regulatory policies.

Sotte, Esposti and Coderoni describe the contributions of agricultural economists based at the Faculty of Economics of UNIVPM. Their focus was on the transformation of the primary sector in Italy and in the Marche region, with a critical perspective towards the prevalent model of development and the role played by the European Union in this context.

The active contribution to the debate on the transition from GDP to alternative measures of well-being is described by Chelli, Ciommi and Gallegati. In line with Fuà's thought, the construction of composite indicators, as well as the application of new statistical techniques, has been proposed to measure well-being at national, regional and local levels.

From the law side, Mantucci, Di Stasi, Torsello, Giuliani, Trucchia, Zarro, Zuccarino, Putti, Calamita and Califano present the research stream on arbitration taking into consideration the positive effects that the spread of arbitration would have on society, growth and competitiveness of Italian companies.

Productivity differentiation, international specialization patterns and the organizational model of firms and industrial districts are the topics discussed by Cucculelli, Lo Turco and Tamberi from the empirical micro-level perspective adopted by the scholars of the Ancona school of economics.

Ballestra, Brianzoni, Colucci, Guerrini, Pacelli and Radi contribute to the studies on quantitative methods in economic and finance bringing two different perspectives: from one side, they analyse the development of new theoretical models and new numerical and analytical methods for computing the default risk, for pricing financial derivatives and for conducting empirical tests to validate the models and, from the other side, they deal with nonlinear dynamic systems in both discrete and continuous time for studying important stylized facts in economics and finance.

Sustainability of real and financial markets is the topic of the paper presented by Manelli, Branciari, Montanini and D'Andrea. The authors contribute to the debate on the agricultural commodity prices, and they present a preliminary analysis on the impact of the financial derivative instruments.

The contribution of Chiucchi, Giuliani, Poli, Gatti, Montemari and Del Bene sheds light on the evolution of the financial and non-financial disclosure particularly focusing attention on the quality of the disclosure of intangible assets. The aim of the chapter is to try to understand the possible evolution of this field of research also by considering the recent introduction of the integrated reporting framework.

De Angelis, Fiordiponti, Giorgini, Lucarelli, Mazzoli and Perna present an interesting research on the impact of the information and communication technology in finance and law; Fintech and Big Data are changing the shapes of the financial industry, and the legal and economic questions deriving therefrom need to be answered.

Following the same vein, Pascucci, Temperini, Marinelli and Marcone describe how SMEs benefit from digitalization in facing the intense global competition, finding market opportunities and developing profitable relationships in business-to-business markets.

Ancona, Italy   Sauro Longhi
                                                                  Andrea Monteriù
                                                                 Alessandro Freddi
                                                                    Giulia Bettin
                                                                  Silvio Cardinali
                                                            Maria Serena Chiucchi
                                                                 Marco Gallegati

Contents

50 Years of Teaching and Research Trends at the Faculty of Economics of the Università Politecnica delle Marche

Francesco Maria Chelli, Gian Luca Gregori, Stefano Marasca and Luca Papi

Abstract This section aims to provide an overview of the research and teaching activities carried out at the Faculty of Economics of the Università Politecnica delle Marche over the last 50 years. In particular, its objective is to show the development of subjects taught and researched with regard to the four main areas of the Faculty: Business Administration, Economics and Business Management, Economic Statistics and Economics. Besides showing how the main areas of research and education have changed over time, this section also aims to stimulate reflections on the future of these areas, shedding light on new streams on which research activities could be focused. In doing so, this chapter combines the past, the present and the future of the Faculty of Economics of the Università Politecnica delle Marche and it shows how changes which have characterised the national and the international context over the last 50 years affected both the teaching and the research activities carried out by the Faculty and the contribution that the latter aims to provide over the next few years.

F. M. Chelli · L. Papi
Department of Economics and Social Sciences, Università Politecnica delle Marche, Ancona, Italy
e-mail: f.chelli@univpm.it

L. Papi
e-mail: l.papi@univpm.it

G. L. Gregori · S. Marasca (✉)
Department of Management, Università Politecnica delle Marche, Ancona, Italy
e-mail: s.marasca@univpm.it

G. L. Gregori
e-mail: g.gregori@univpm.it

© Springer Nature Switzerland AG 2019
S. Longhi et al. (eds.), *The First Outstanding 50 Years
of "Università Politecnica delle Marche"*,
https://doi.org/10.1007/978-3-030-33879-4_1

1 Teaching and Research Trends in Economic Statistics

In our department, the field of economic statistics (SECS-S/03) has undergone considerable changes in the last 40 years. Originally, there were two main statistics courses: Statistics, as in methodological issues, and Economic Statistics, tied to matters dealing with statistical methodologies and models and the processing of data and indicators.

Today these two main categories have fragmented into nine different courses: Statistics 1 and 2 and Multivariate Statistics with R, which belong to Statistics, in its methodological meaning; Actuarial Statistics, Business Statistics (Italian & English courses), Economic Statistics, Statistics for Financial Markets and Statistics for Business Decisions, which belong to the category of Economic Statistics. This evolution also involves research carried out by the faculty, starting with contributions by professor Ornello Vitali in the 1970s.

The original formation of the Economic Statistics research group dates back to prof. Ornello Vitali, and in particular to his attendance in the program for economic development organized by the Social Science Research Council; coordination of the Italian group was attributed to Giorgio Fuà and general coordinators M. Abramovitz and S. Kuznets. Relevant contributions, the result of decades of study, are represented by the reconstruction of the active population based on a deep, annotated work. The results constitute the basis for the critical reconstruction and interpretation of the Italian economy's developmental process with reference to the service branch.

Relevant theoretical and empirical contributions have been made in the field of territorial statistics, which have led to a substantial growth of the discipline. They have also brought a radical change to studies with the introduction of discriminant analysis to identify rural and urban areas, going beyond the rural–urban dichotomy and considering the continuum for the first time, developing a new way of thinking that is still relevant (Merlini 1978; Vitali 1983). By studying several stochastic models, some relevant internal migration dynamics are gathering attention (Mattioli and Merlini 1983).

The field of research explored in the eighties and nineties relates to the functional distances through an initial work (Chelli et al. 1988a) that aims to delimit areas by using both functional and geographical areas and substitutes classical zoning algorithms using discriminant analysis. Another paper (Chelli et al. 1988b) analyses functional regions obtained by disaggregating the matrix of migration flows in the Province of Catania according to economic sector. It also illustrates a comparison of the four different regionalizations (agriculture, industry, commerce, and other activities) and a classification obtained following the areal typologies approach. Another work worthy of mention (Chelli et al. 1988c) describes the use of functional distances to delimit metropolitan regions. The results of the experiments are shown for the Strait area using the Brown–Holmes algorithm and an algorithm proposed by the authors. The latest results are more effective in cases where it is necessary to determine separate broad regions.

Finally, the last research in this area (Chelli et al. 1992) develops a first attempt at the joint analysis of urbanization that combines the two classical components of this kind of study: the identification of functional regions and areal typologies.

In 2000, attention was brought to price dynamics, with the estimation of consumer price indexes (Chelli and Mattioli 2007). In this work, consumer price indexes for the entire country and those for individual households are weighted arithmetic averages of relative prices, which differ essentially in terms of their weighting systems. Whereas the former use proportions of total expenditure on goods and services, the latter use proportions of expenditure by each household. In the usual calculation of the country-wide index, each household contributes to determining the national index with a weight proportional to its expenditure. In other words, households that spend more—that is, wealthier ones—are represented in the calculation of the national index with a greater weight, which explains why it is termed the plutocratic index. In contrast, in *democratic indexes*, the same weight is assigned to each household. This paper presents an initial estimation of democratic price indexes for Italy in the period 1995–2002. The results are extremely interesting in that they highlight significant differences between the two calculation methods. A distinction between *democratic* and *plutocratic prices* was shown in Chelli and Mattioli (2008). The paper presents a first estimation of the democratic price indexes for Italy in the period 1995–2002 and for Marche Region in the period 1995–2005. The two methods, *democratic* and *plutocratic*, for calculating the national index, which are based on different weighting systems, lead essentially to results, invariant with respect to the territorial definition and of the considered level of aggregation.

From prices, the attention shifts to expenditure in Italy (Chelli et al. 2009). This paper explores the determinants of heterogeneity in the expenditure behavior of Italian households, using the Households Expenditure Survey provided by the Italian National Institute of Statistics (ISTAT) for the year 2005. Authors assume that differences among consumers are associated with differences in their economic and socio-demographic characteristics (such as gender, employment status, and age of the householder, number of household components, presence of components under 18 years), and they look for those characteristics that better differentiate groups of households according to their purchasing patterns. They apply a *nonparametric discriminant analysis* based on the various expenditure budget components, and detect the most discriminating partitions of families. This technique also allows them to identify the specific consumer goods that differ significantly across the groups identified by the best partitions. They then study the different effects of price dynamics on subgroups of households, and propose consumer price indices specific for the optimal households groups. Another relevant work measures how consumption distribution was polarized among Italian families during the years 1997–2006 (Chelli et al. 2010). Since a polarization measurement requires a precise definition of groups, they consider the household partition that best explains the variability of households in their purchasing patterns. In particular, authors use a *discriminant analysis* technique to examine the determinants of the heterogeneous expenditure patterns of Italian households, using the Households Expenditure Survey provided by the Italian National Institute of Statistics.

Another field of research we have examined is index numbers (Mattioli and Lamonica 2010). This paper presents an empirical analysis of the response of several methods proposed in the literature to rank the alternatives, determine their weights, and perform accurate comparisons among economic phenomena. The various scaling methods are evaluated using real data provided by the Italian National Institute of Statistics (ISTAT). The most relevant results are the close agreement of the weights obtained with the various methods and the robustness of the evaluation.

Finally, economic statistics research in this period evolved to include *interregional input–output tables* (Mattioli and Lamonica 2012). The fundamental information used in input–output tables concerns the flows of products from each industrial sector, considered producers, to each of the sectors, both itself and others, considered consumers (Lamonica and Chelli 2018). This basic information from which an input–output model is developed is contained in an interindustry transactions table. The rows of such a table describe the distribution of a producer's output throughout the economy (Bonfiglio and Chelli 2008). The original Leontief model is set at the national level, and does not disaggregate transactions between and within subnational regions (Mattioli and Lamonica 2015a). Interest in the subnational economy has led to the introduction of a regional dimension in the input–output framework (Mattioli and Lamonica 2015b). The fundamental contribution provided by the group of economic statisticians in the analysis of interregional input–output tables is related to the performance of various non-survey techniques for constructing sub-territorial input–output tables and the empirical analysis of the world's economic structure, the degree of interaction between the economic systems of countries, and their evolution over time (Mattioli and Lamonica 2015c).

Some statistics and demographic contributions emerge from Lamonica and Zagaglia (2013). In this paper, authors studied the determinants of internal migration in Italy from 1995 to 2006. To conduct this investigation, they applied an augmented version of the gravity model to migratory flows of Italians and resident foreigners. In addition to the classic determinants of migration—i.e., population sizes and the distance between places—the model considers a possible autocorrelation of flows and a set of socioeconomic and demographic explanatory variables that may influence migratory flows. Different results were obtained for the two subpopulations. Among the Italians studied, both economic conditions and regional demographic features were found to operate as both push and pull determinants of migratory flows, although the demographic characteristics were shown to have affected migratory flows to a lesser extent. Among the resident foreigners studied, the demographic characteristics of the regions did not appear to have acted as push factors, but they were found to have an effect as a pull determinant. While the economic conditions of the destination regions were shown to have been particularly important in attracting the resident foreigners, the economic conditions of the sending regions were not found to have had a clear-cut effect on the decision to leave.

Very recently, our research has focused on new objectives that are still in line with Vitali's since they aim to solve practical social/financial/economic problems by investigating social/financial/economic behavior. This is done by inferring the behavior from its effects on financial and social dynamics. The idea of formulating *behavioural* statistical models falls in line with Fuà's teachings, the agent-based school of thought developed in the Faculty of Economics in the last 10 years, as well recent lines of research concerning intangibles. Specifically, the main research lines focus on two main areas. The first is the analysis of welfare dynamics with specific attention to the definition of well-being and its measurement. This research has embraced a multidimensional perspective in the study of well-being, promoting and actively contributing to the debate on *beyond GDP*, with a focus on two key aspects: (i) The space dimension, by studying Italian well-being on the national (Chelli et al. 2016), regional (Chelli et al. 2018; Gigliarano et al. 2014), and local levels (Chelli et al. 2016; Ciommi et al. 2017a); and (ii) The time dimension, by proposing a measurement of well-being in the long (Chelli et al. 2018; Ciommi et al. 2017a) and short terms (Ciommi et al. 2017b).

The second line of research relates to the analysis of financial market dynamics by using behavioral models, that is, models with parameters capable of interpreting agents' beliefs. Panel data analysis is used to obtain time series of the estimated parameters, which are used to interpret the changes in investors' moods in terms of risk aversion and risk premium (Mariani et al. 2018) and to detect financial institutions' instabilities/strategies (Tedeschi et al. 2018; Recchioni et al. 2017a, b; Recchioni and Sun 2016).

2 Teaching and Research Trends in Economic and Business Management

During the last 50 years, our faculty has witnessed a considerable growth of Economics and Business Management (SECS-P/08) related teachings, both from a quantitative and qualitative perspective.

From a quantitative perspective, the numerical growth of teachings within the degree courses is clearly visible in Fig. 1. Until the late 90s there were solely two teachings related to our disciplinary grouping: *Tecnica Industriale e Commerciale* (literally, Industrial and Commercial Technique), dealing with topics related to all the

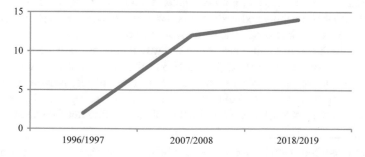

Fig. 1 Evolution of SECS-P/08 related teachings

functional areas of the company (from production to marketing), as well as strategic and organizational problems; and *Tecnica del Commercio Internazionale* (literally, International Commerce Technique), dealing with international marketing topics.

During the following years, the university teaching programs broadened considerably, so to reach 14 Economics and Business Management related teachings in the academic year 2018/2019, three of which are in English. Among these, marketing teachings have accelerated in a significant way, also coherently with the national scenario and considering the evolution of skills required in the job market.

From a qualitative perspective, the growth in teachings' heterogeneity with increasing specialization levels is worth noting. Alongside specific teachings related to the *traditional* subjects focused on functional areas (e.g. Production and Logistic, Marketing, International Marketing, Sales Management, etc.), additional specialized subjects have been introduced, whereby lecturers have also prompted specific research paths (e.g. Internet and Marketing, Business Marketing), and teachings focused on specific types of companies (e.g. Economics and Business Management of Tourism Companies, Economics and Business Management of Trading Companies). [1]

The teachings evolution of the last years has mirrored the evolution of the research topics of the Economics and Business Management research group, with a focus on Marketing and Operations (Silvestrelli and Bellagamba 2017). Adopting mainly a qualitative methodology, based on in-depth case studies and personal interviews with managers, entrepreneurs and trade associations (Runfola et al. 2017), the studies have regarded different topics; however, *price* and *distribution channels* have been the two main investigated marketing mix elements. Some interesting contributions have deepened the relationships between industrial firms and commercial firms, with the aim of understanding the implications of the distribution system evolution on the marketing strategies of industrial firms (Gregori 1995) and on the sales management and organization (Perna et al. 2013). The growing strategic role of commercial firms in the economic landscape, as a consequence of the *commercial revolution* related to the coming of *mass consumption*, has transformed the relationships between industrial and commercial firms; a complex distribution network made up of manufacturers, wholesalers, retailers and transportation channel specialists, all of whom collectively enabled goods to flow to consumers, was established. Among these actors, complicated relationships developed, both competitive and collaborative. This revolution drastically changed established methods of marketing and distributing both goods and services. New retail and wholesale formats had a profound impact on the economy of the twentieth century and transformed the way that consumers purchased goods and services. More recently, the exponential rise of e-commerce, as a new electronic distribution channel, has further modified the structure of distribution system; this phenomenon provides a myriad of opportunities for conducting both theoretical and empirical future research, for example on the role of existing middlemen and the role

---

[1] It is worth remembering that in the past, eminent scholars have taught in our faculty, such as Aldo Burresi, Gennaro Cuomo, Giorgio Eminente, Gaetano M. Golinelli, Sergio Silvestrelli and Riccardo Varaldo.

of new electronic intermediaries. In particular, small and medium-sized manufacturers could benefit from e-commerce in a significant way, expanding their geographical scope and establishing new and direct relationships with customers (Gregori et al. 2016). However, Small and Medium-sized Enterprises (SMEs) are lagging behind in e-commerce (and, more in general, in ICT) adoption (Pascucci et al. 2017). This is a serious issue, considering that SMEs represent the backbone of the European economy and they play a key role in national economies around the world. According to this, great attention was paid by the research group to studies concerning small and medium-sized firms; these studies have been often aimed to understand how SMEs could cope with the emergent challenges of globalization and, more recently, of digitalization (Pascucci and Temperini 2017).

It is worth noting that a significant part of the empirical researches carried out by the research group is focused on a specific industry, such as the furniture industry (Silvestrelli 1979), footwear industry (Gregori et al. 2012), wellbeing industry (Gregori and Cardinali 2015), coffee industry (Pascucci 2018; Giuli and Pascucci 2014), knitwear industry (Marcone 2013), integrated circuits industry (Marcone 2004), kitchen hoods industry (Bellagamba 2018). Particular attention was given to Made in Italy and to the problems of counterfeiting and Italian sounding (Gregori 2016). These studies have allowed deepening the knowledge of the dynamics and the relationships in specific contexts, taking in consideration that each industry shows the own strategic, operative, organisational characteristics.

A common trend among all these contexts is the relevance of inter-organisational relationships.

In the Marketing field the relationship and network perspectives have become increasingly important (Ferrero 1992; Hogan et al. 2002): the first one focuses on the supplier–customer relationship, which is a sustainable source of competitiveness and profitability (Gummesson 1997; Grönroos 2004; Perna et al. 2015); the second one focuses on inter-organizational relationships, network structures and network types (Möller 2013).

Relationship marketing first developed in the BtoB and service markets, however, following the digital revolution and Web 2.0 developments, which facilitate interactions between businesses and consumers and among consumers, now it can also be applied in BtoC markets (Pascucci 2013). From this standpoint, a close connection exists between the opportunities created by the digital and the relational perspective, typical of a marketing service-dominant logic (Vargo and Lusch 2004, 2008), rather than good-dominant. The use of new communication media (particularly social networks) offers companies the opportunity to intensify and deepen their interactions with customers, laying the groundwork for the development of medium and long term trust relationships, which can result in a real *customer engagement* (Youssef et al. 2018).

Therefore, from a thematic perspective, *relationships, knowledge* and *digital technologies* are the keywords inspiring the current marketing teachings' programs and marketing research, to emphasize the increasingly important role of intangible resources (intangibles) (Marchi and Marasca 2010) in the economy and in business management.

It is not surprising that the Department of Management has also grounded on these keywords to build the project *Dipartimenti di Eccellenza 2017* (literally, Outstanding Departments 2017), awarded by MIUR in 2018. The focus of the project on intangible resources arises from the tremendous changes resulting from the *digital revolution*, which is foreshadowing a new industrial era, also labelled as Fourth Industrial Revolution (Schwab 2015). The phenomenon *Industry 4.0* implies a radical reconfiguration of production and distribution processes, in both real and financial markets, making *immateriality* increasingly pervasive (Bellagamba et al. 2018; Arnaboldi et al. 2017). Finally, it should be noted that the trend towards increasingly data-driven marketing, fueled by big data from multiple and heterogeneous sources (Mandelli 2017), will impose in the near future to invest resources also in university teaching programs aimed at developing skills related to the management and the analysis of (quantitative and qualitative) data for corporate decision-making processes. Marketing analytics is an important skill for marketing students and it is urgent for schools to design marketing analytics courses in order to accommodate market demand for these new skills (Liu and Burns 2018). Since 2000 the integration of quantitative methods in marketing education textbooks and teaching practices has been a quite evident trend, which was facilitated by the availability of much more data thanks to the digital revolution (Ferrel et al. 2015). Big data is "the new capital in today hyper-competitive marketplace" (Mayer-Schönberg and Cukier 2013); however, many firms fail to exploit its benefits, because of the complexity in the process of converting Big Data in marketing insights first, and marketing decisions then: this is one of the most important challenges recognized by marketing executives (Leeflang et al. 2014). The main obstacle is the shortage of data scientists, as marketing departments in business schools have been slow to design curricula to generate such skills (Erevelles et al. 2016). As a consequence, this *digital talent gap* represents a big challenge also for business school, which should develop a strong focus on analytical skill and the use of data in marketing decision making (Liu and Burns 2018; Leeflang et al. 2014). These are transversal skills that require a systemic and integrated approach in the design of effective university teaching programs, where marketing (and more in general management) studies have to be integrated with Statistics, Economics and Management Accounting. The analytical and quantitative skills are not enough; they should be complemented with a strong background in business disciplines.

3   Teaching and Research Trends in Business Administration

Research and teaching activities carried out by the Department of Management in the field of Business Administration—*Economia Aziendale* (SECS-P/07)—over the last 50 years have evolved significantly. This development can be largely traced back to changes in both the national and the international scenario at normative and competitive levels.

With regard to teaching, in the 1980s financial accounting was the object of the two main courses offered in Business Administration at the Università Politecnica delle Marche: *Ragioneria I*, which covered financial accounting, and *Ragioneria II*, which dealt with extraordinary corporate transactions.[2] Starting in the 90s, attention paid to these issues gradually increased, leading to seven courses being offered by the end of the decade and twelve by the end of 2000. Over the last ten years, the number of courses taught in this field has increased to fourteen and they are now offered both in Italian and in English. Nevertheless, what should be noted is that, in the wake of the evolution of Business Administration, there has also been a gradual growth in the variety of subjects offered over time.

With regard to research, in the early stages it predominantly concerned financial accounting. The lack of specific and complete regulations concerning financial statements made necessary an exploration of how company economic results, assets, credits and liabilities could be measured, evaluated, represented and communicated. This initial focus derived from the need to fill a practice gap that had been caused by the absence of laws aimed at regulating the main objectives of the financial statement and the way it should be prepared. Therefore, the contribution made by research and the academic debate was fundamental in this stage because of the abovementioned lack of laws for regulating a relatively new field.

It was through Law no. 216/74 that the abovementioned gap was partially filled also, and especially, thanks to the contributions of research carried out during the previous stage. The Law established the structure of the balance sheet and of the income statement. This attempt to introduce a set of rules aimed at standardizing the preparation of the financial statement also favored the rise of new research avenues which were linked to the auditing activities that Law no. 216/74 required in order to ascertain financial statement preparers' degree of compliance with what the Law itself established (Branciari 2000).

Nevertheless, the introduction of Law no. 127/91, which transposed the IV and V Directive of the European Commission in Italy, essentially rewrote the laws which, up to that moment, had regulated the financial statement. Moreover, it was during this time period that the *Economia Aziendale* research group was established at the Università Politecnica delle Marche, thanks to the fundamental impetus provided by Prof. Luciano Marchi.

The establishment of a set of principles and rules and of a fixed and very detailed structure of the balance sheet and of the income statement led to a deep change in research interests (Marchi and Marasca 1994). Attention shifted to the evaluation processes that the new reform made necessary. In fact, they were no longer understood as a means to *mystify* reality but rather, as an unavoidable practice in a normative context in which legislators require preparers to evaluate in a discretional way some

---

[2]It should be underlined that some of the foremost Business Administration scholars, like Vittorio Coda, Antonio Tessitore and Isabella Marchini, taught at the Faculty of Economics of the Università Politecnica delle Marche.

of the most relevant items represented in the financial statement (Marasca 1999). In the meantime, a stream of research was also devoted to the analysis that the new financial statement allowed, namely of the set of indicators and information that can and should be communicated to internal and, in particular, to external stakeholders. This research interest stemmed from the perception that a trade-off was gradually emerging between the quantity of information requested by financial analysts and other external stakeholders, and its actual quality and relevance.

It should be argued that this stage of the evolution of legislation on the financial statement opened up important avenues of research in which the Department of Management was deeply involved. One of these was focused on the analysis of the financial statement and of its disclosure function to external stakeholders in different contexts, like that of financial intermediaries (Branciari 2003), of small enterprises (Branciari 2011) or of company groups (Marchi et al. 2010). Another significant research path concerned the adoption of national and international accounting standards and, in particular, their effects on the comparability of financial statements (Branciari and Poli 2009).

The shift in the research activity in the Department of Management from one predominantly focused on financial accounting to a new stage is in part also attributable to the widespread international diffusion of studies carried out by Harvard Business School scholars. In particular, the seminal work by Robert Anthony "Planning and control systems: A framework for analysis", published in 1965 and translated into Italian in 1967, shifted the attention of a large number of scholars from financial to managerial accounting., e.g. management accounting. This should be understood as something more than a mere change of research interests. Rather, it was the consequence of adopting a different way of understanding and interpreting companies and the function that accounting systems can play within them. Starting from the work by Anthony, performance was no longer understood only in a unitary way, namely at the company level. It began to be considered, instead, as the consequence of the analytical contribution given by every organizational unit within the company. According to this view, performance measurement should entail the adoption of indicators aimed to provide both an overall picture of the company performance and a detailed picture of the results achieved by each organizational unit. In addition to this, to ensure the use of information for decision-making purposes, the accounting system had to move from the provision of extremely accurate information based on the results achieved to the production of timely and forward-looking information. In other words, from a management accounting perspective, the internal use of information provided by the accounting systems started to gain relevance since management accounting systems were, and still are, primarily seen as systems aimed to support managerial decision-making processes rather than external stakeholder's evaluation processes.

Contributions in the field of management accounting have covered a wide range of topics. Particular attention was paid to the analysis of the design as well as the functioning of management accounting systems in specific contexts, like that of small and Medium Sized Enterprises (SMEs) (Branciari 1996; Marasca 1993) or that of trading companies (Marasca 1989). Moreover, research has also been directed to explore the tools that help management accounting systems work and, in particular, their degree of implementation at a national level and with regard to specific sectors (Marchi et al. 2009; Marasca 1991, 2003).

It should be argued that the journey outlined above has taken a very different course for research in the field of public and health-care organizations. In these contexts, in fact, research on the design and the implementation of management accounting systems originally attracted the attention of scholars because of the great normative pressure to measure both the effectiveness and the efficiency of their internal processes. It has only been in recent times, however, that research has started to focus attention on financial accounting systems in these organizations and this is predominantly due to the radical normative changes that forced public and health-care organizations to shift from the adoption of cash accounting systems to the use of accrual accounting systems. This has opened a very prolific avenue of research concerning the main cultural and organizational barriers that can arise when such a complex process of transition needs to be implemented in these peculiar contexts.

In particular, research carried out by the Department of Management has explored in depth the contribution that management accounting systems can give to the management of public and health-care organizations. At least in the initial stages, particular attention was devoted to the technical aspects related to the adoption and use of management accounting tools in these specific contexts and to the levers that could facilitate their implementation (Marasca et al. 2013; Del Bene 2000, 2008; Marasca 1998). Gradually, the issues under investigation broadened, and attention shifted from the analysis of management accounting tools to barriers which can hinder their adoption, making them ineffective, in the context of public and health-care organizations (Del Bene and Russo 2017; Del Bene 2014; 2016; Marasca 2001). Moreover, research also focused on the way management accounting tools can be useful in these contexts in order to improve managerial decision-making processes and support managers in achieving higher level of efficiency (Marasca 2016; Marasca et al. 2016). Recent research has also investigated how advanced management control tools, such as those related to so-called strategic control systems, could be adopted, and the effects they can lead to, in the context of public administrations.

A third stage of the research activity carried out by the Department of Management in the field of Business Administration is closely tied to the rise, in recent years, of new business paradigms. The acknowledgement that the competitive advantage is more and more frequently linked to firm-specific and intangible assets, e.g. knowledge and trust, led scholars to a profound re-orientation of their research. This prompted the Department of Management to activate a new and very prolific avenue of research focused on these new critical assets, ranging from their measurement (Marasca et al. 2009), to their evaluation (Giuliani and Marasca 2011, 2018), to the use of information related to them for managerial purposes (Chiucchi et al. 2014, 2016, 2018).

The relevance gained by these new assets has deeply affected both financial and management accounting systems. Financial accounting systems, in fact, stem from the acknowledgement that new information needs to be communicated to external stakeholders since that is how they can actually know the *real* value of a company. Management accounting systems, instead, have profoundly changed the way company performance is measured and the type of information that is provided to managers in order to support their decision-making processes. Attention has shifted from traditional financial information to non-financial indicators and from information focused on the efficiency of production processes to information focused on new critical and strategic issues like customers, suppliers, partners and competitors. This change has also stimulated a strong cooperation between accounting scholars and scholars in other disciplines, like marketing or finance.

The acknowledgement that cross-fertilization among different disciplines is an unavoidable necessity has found its concrete realization in the research project proposed by the Department of Management and awarded by the MIUR (Italian Ministry of Education, Universities, and Research) in 2018. The idea to create a laboratory for digital marketing and business intelligence finds its roots in the awareness that new research avenues will be based on the cooperation between different areas and fields. This is the case for digital accounting which will require strict collaboration between accounting and marketing disciplines to identify, select, and present information obtained through new data sources, e.g. big data, and to make it available and useful to internal users. Similarly, the complexity of the sources from which this new information can be obtained make it necessary to establish tightly-knit cooperation between accounting and IT disciplines in order to identify those channels through which information can move from the producer to the final users in a context in which timeliness becomes, evermore frequently, a critical success factor for companies.

These are some of the new challenges that arise for Business Administration scholars who are called to reflect on the consequences that these outlined changes in the overall scenario can have on the production and the use of information for internal and external users. On one side of the issue, the abovementioned evolutions will entail a stronger focus on qualitative methods, especially the case study method, to provide a deep exploration of the way they can influence the design as well as the functioning of financial accounting and managerial control systems. On the other side, the rise of digitalization and the opportunities ensuing from the spread of business analytics and artificial intelligence will make it necessary to establish a deep cooperation among scholars who are used to working in different fields of research, from statistics to management, from marketing to banking and finance. In fact, these new phenomena will change the role and the functions that financial accounting and managerial control systems can play within organizations. Therefore, their analysis can be relevant in the effort to develop new theoretical frameworks and to support, from a scientific standpoint, the rise of new theories able to interpret their effects.

4 Teaching and Research Trends in Economics

In recent years the economists of the Faculty of Economics "Giorgio Fuà" have carried out intensive research in close interaction with their teaching activity. The latter has given rise, over the years, to a large number of degree programs focused on economic disciplines. Particular mention should be made of the very early establishment (1985) of the Ph.D program in economics which has been among the first Italian doctorates to be activated.[3]

Presenting here some of the main strands of the research carried out by the Ancona economists, we avoid recalling the quantitative aspects that would lead us to indicate a considerable number of papers, produced by a conspicuous number of scholars, on a wide range of economic themes. We do not have space here neither to list individual themes, nor all authors' contributions.[4] We limit ourselves to identifying some of the key words of the Ancona approach to the analysis of the economy. Then we will analyse in more details how these key words have been declined within the area of monetary studies recalling a few examples. Some other contributions on some other main economic research areas are instead reviewed in dedicated chapters of this volume.[5]

In the first decades of the life of the Faculty of Economics, the most studied issues by the economists of the Ancona group can be identified in the analysis of economic development applied mainly to Italy and to those countries that were labelled by Giorgio Fuà "countries with recent development". At that time, attention to development meant analysing real sector issues. According to the prevailing literature, the main determinants of the economic development were identified in variables of a real nature that mainly recalled aspects such as the role of savings, capital, infrastructures, human capital, institutions etc. In that initial phase, there was not much room for financial variables. At that time, in Ancona, the most important lines of research followed the intuitions, ideas and assumptions of Giorgio Fuà, reviving the influential contributions of scholars like Schumpeter, Kuznets and Abramovitz. A common factor was the endogenous character of economic development that for each situation reflects several variables such as different historical roots, diversified social capability and various organizational and entrepreneurial factors (Fuà 1980). Inspired by those beliefs, the theme of economic development has thus been declined in several different but interconnected lines of research. The Ancona approach has given rise to many contributions in many economic areas that range from the unification of Italy

---

[3]The history of the Italian doctorate is relatively recent. If in some European countries this program had been present for many decades, in Italy it was established only in 1980. In fact, it was activated in 1983. See Ermini et al. (2016) for an analysis of the experience and history of the Ancona doctorate in economics.

[4]For a reference point representative of the first period of research of the Ancona economic school, see the volume "Trasformazioni dell'economia e della società italiana" (Gruppo di Ancona 1999).

[5]See Cucculelli, Lo Turco and Tamberi "Productivity differentiation and international specialization", Palestrini and Russo "The agent-based models school in Ancona", Sotte Esposti and Coderoni "Destination Europe: the transformation of agriculture between decline and renaissance" and Chelli, Ciommi and Gallegati "From GDP to BES: the evolution of well-being measurement" in this volume.

to the European integration. As examples of such richness of topics we can recall the analysis of the labour market and the role of innovations to promote economic growth (Sterlacchini 2008), the productivity gap between and within countries (Conti and Modiano 2012), the relevance of demographical aspects (Fuà 1986), the focus on the development of those Italian regions where the local economic systems were mainly based on localized "industrial districts" of specialized small and medium size enterprises (the North East and Central part of Italy).[6]. Another line of research was that of economic development and European integration studied in the perspective of major structural and long-term transformations (see, among others, Alessandrini and Conti 1981). A further strand of research focused on the territorial analysis with the study of the causes and policies for local development (see, among others, Fuà 1991).

Some of the key words of the Ancona school can be identified as *structural aspects, economic development, long run*. But these words, absolutely recurrent in the economic studies carried out in Ancona, are also traceable in the works of a small group of *Ancona monetary economists* who turned their analyses towards financial issues. In this regard, with reference to this first phase of life of the Faculty of Economics, the works of economists such as Fausto Vicarelli, Giacomo Vaciago, Alberto Niccoli and Piero Alessandrini come to mind. The starting point was due to the contributions and ideas of Fausto Vicarelli, who in turn referred to the pre-Hicksian lines of inquiry (especially those of Schumpeter and Hilferding) and was strongly based on the firm belief that the real and financial aspects of economic development are closely interrelated.

Among the different channels of conjunction between the real and financial spheres analysed by the group of Ancona we can mention two examples of the first period examined here. As first example, we refer to the theme of accumulation of financial wealth (see the works contained in Niccoli 1989) where special attention was paid to the consequences of excessive financialization and excessive accumulation of gross financial wealth. Too much finance involves a higher probability of crisis and therefore concerns about the stability of the system. Moreover, changes in wealth are not only affected by the savings accumulated via income, but increasingly by changes in the capital account. In other words, the formation of wealth is not so much the result of phenomena related to flows (incomes and profits) but beyond a certain point it is largely due to the management of real and financial stocks. This wealth dynamic affects the income distribution that tends to polarize favouring the richest and those groups most equipped for wealth management. Instability and a more concentrated income distribution also harm economic growth over the long term. On the other hand, it is no coincidence that the most recent empirical evidence between finance and growth questions the existence of a monotonous relationship between the financial system and growth, but on the contrary that relationship becomes more dubious when financialization reaches very high levels.

---

[6]The literature on industrial districts is very wide and several contributions have been provided by the Ancona school; see, among others, Fuà and Zacchia (1983), Crivellini and Pettenati (1989)

A second research topic of the first period had investigated the structural and long-term impact of a high public debt shedding some lights on some original and little studied aspects related to a very high public debt situation. A maieutic function of the growth and diffusion of public securities among economic agents is underlined, as well as the effects that the rise of public securities have had on banks behaviour. More in general the presence of an high public debt can facilitate financial asset management (a growing role of financial intermediaries in addition to that of transferring flows) arriving in this way to identify a positive structural effect of the diffusion of public debt on the efficiency of the whole financial system (Niccoli and Papi 1993). Of course, concerns about the health of public finances are not lacking, but even when the group of Ancona discusses the negative consequences of the public debt, the angle of analysis is not only the usual one of macroeconomic compatibility, but again that of structural issues and the changing behaviour of economic agents. So, for instance, the high level of public debt does not bring out much for the amount of the debt as for its burden due to the enormous flows of interests that affect the income redistribution - from the most productive classes to other sectors. All this strengthens and nurtures the typical mentality of the rentier focused on income and speculative operations rather than on productive and entrepreneurial logics. In other words, the risk is that instead of the well-known euthanasia of the rentier, the euthanasia of the productive classes (entrepreneurs and workers) is encouraged with all the relative consequences on the potential development of the economic system.

Since the early nineties, the research of Ancona in the monetary and financial field has been extended on other fronts with studies that have covered topics such as monetary policy, national and international financial integration, bank crises, the role of local banks and the bank-firm relationship. In the meantime, new human resources have fueled the founding group of Ancona's monetary studies. In the 1990s, new economists, among whom we remember Luca Papi, Alberto Zazzaro and Andrea Presbitero and Michele Fratianni who joined later the team of Ancona. And with the enlargement of the human resources, the contributions and the strands of research also widened. We only mention a few here. As of European integration, Monticelli and Papi (1996) propose an innovative perspective on the issue of improving the European monetary integration. Their work identifies the economic conditions underlying the choice of the optimal scheme of monetary coordination under an exchange rate agreement and shows that one of the key conditions is the relative stability of the European Union area-wide demand for money. On this basis the authors provide the first exploration of the properties of the demand for money for the European Union as a whole. Relevant methodological and empirical contributions have concerned the issue of the efficiency of the banking system (Favero and Papi 1995); in this field Lucchetti et al. (2001) provide a new way of analysing the relationship between banking and economic growth by applying an innovative indicator of the development of the local banking sector based on a measure of bank microeconomic efficiency. Since 2007—when the Money and Finance Research group (MoFiR) was established within the Department of Economics by a small bunch of monetary economists—many others contributions have been produced by the Ancona monetary group. However, despite the wide spectrum of papers published, they still share a common feature: all themes are analysed in a perspective of structural and long-term evolution. Within the various works, the themes of the economic development have remained central, with special attention to the role of some heterogeneities on the growth of territories. The basic hypothesis is that the initial territorial conditions are not levelled and that this requires a dedicated approach also for the banking and financial field. In this way the financial system could contribute to the reduction of development gaps at regional levels. In particular, the contributions on the efficiency and geographical organization of banking systems and their impact on local growth are noteworthy. On these topics there is an interesting and intense debate driven by the structural transformations of banking systems due to technological and regulatory innovations. The prevailing opinion in the literature is that technological advances and financial innovations would led to an increasingly transaction-oriented banking industry and that this would lead to a standardization of banking products and a concentration of structures with an economy served only by few major international players. Consequently, the importance of local banks and geographical proximity between banks and borrowers would be doomed to decline over time. Within the Ancona group some critical voices warned against the myth of global banking, pointing out the role of banks as agents of development, and the high costs that the consolidation and geographical agglomeration of banks would generate for peripheral territories and small local firms (Alessandrini and Zazzaro 1999). This line of research has been very productive within the Ancona group and it is reviewed on a dedicated chapter of this book.

References

Alessandrini, P., & Conti, G. (1981). *Commercio estero e allargamento della CEE. Prospettive per l'industria italiana*. Bologna: Il Mulino.

Alessandrini, P., & Zazzaro, A. (1999). A "possibilist" approach to local financial systems and regional development: The Italian experience. In R. Martin (Ed.), *Money and the space economy* (pp. 71–92). New York: Wiley.

Anthony, R. N. (1965). *Planning and control systems: A framework for analysis*. Boston: Harvard Business Review Press.

Arnaboldi, M., Busco, C., & Cuganesan, S. (2017). Accounting, accountability, social media and big data: Revolution or hype? *Accounting Auditing & Accountability Journal, 30*(4), 726–776.

Bellagamba, A. (2018). *Sviluppo e ristrutturazione delle imprese produttrici di cappe. Il caso del distretto di Fabriano*. Torino: Giappichelli Editore.

Bellagamba, A., Gregori, G.L., Pascucci, F., Perna, A., & Sabatini, A. (2018). Industria 4.0: non solo una rivoluzione tecnologica. In: Le competenze per costruire il futuro (AA.VV.), Edizioni di comunità, Roma.

Bonfiglio, A., & Chelli, F. M. (2008). Assessing the behavior of non-survey method for constructing regional input–output tables through a Monte Carlo simulation. *Economic Systems Research, 20*(3), 243–258.

Branciari, S. (1996). *I sistemi di controllo nella piccola impresa*. Torino: Giappichelli.

Branciari, S. (2000). Il bilancio falso e inattendibile e il giudizio del revisore contabile: legami e implicazioni. *Rivista dei Dottori Commercialisti, 51*(3), 373–400.

Branciari, S. (2003). *La comunicazione economico-finanziaria degli intermediari finanziari.* Milano: Franco Angeli.

Branciari, S. (2011). Il rendiconto finanziario e le piccole imprese: un modello operativo per i soggetti esterni. *Financial Reporting, 3,* 107–124.

Branciari, S., & Poli, S. (2009). Incomparabilità dei bilanci IAS-IFRS? *Prime riflessioni, Analisi Finanziaria, 74,* 5–19.

Chelli, F. M., Ciommi, M., Emili, A., Gigliarano, C., & Taralli, S. (2016). Assessing the equitable and sustainable well-being of the Italian provinces. *International Journal of Uncertainty, Fuzziness and Knowledge-Based Systems, 24*(Suppl. 1), 39–62.

Chelli, F. M., Ciommi, M., Ermini, B., Gallegati, M., Gentili, A., & Gigliarano, C. (2018). San Matteo e la provvidenza. I luoghi ei tempi dello sviluppo italiano. *Rivista giuridica del Mezzogiorno, 32*(3), 643–672.

Chelli, F. M., Gigliarano, C., & Mattioli, E. (2009). The impact of inflation on heterogeneous groups of households: An application to Italy. *Economics Bulletin, 29*(2), 1297–1316.

Chelli, F. M., Gigliarano, C., & Mattioli, E. (2010). Household expenditure polarization in Italy, 1997–2006. *NIKE 1,* 21–23.

Chelli, F.M., & Mattioli, E. (2007). What consumer price index? A first attempt to estimate a democratic index for Italy 1995–2002. *NIKE* (2–3).

Chelli, F.M., & Mattioli, E. (2008). Indici democratici e plutocratici: la scelta è così rilevante? Uno studio sulla realtà italiana e marchigiana, in Economia Marche (I).

Chelli, F.M., Mattioli, E., & Merlini, A. (1988a). Un approccio ai problemi di zonizzazione mediante l'uso congiunte di distanze funzionali e distanze geografiche: valutazione del loro potere discriminante. In: Proceedings from the Giornate di studio del gruppo italiano degli aderenti all'IFCS, Erice-Trapani (pp. 151–179).

Chelli, F.M., Mattioli, E., Merlini, A. (1988b). Delimitazione delle regioni funzionali distinte per ramo di attivita'. In: Proceedings from the Giornate di studio del gruppo italiano degli aderenti all'IFCS, 181–198, Erice-Trapani.

Chelli, F.M., Mattioli, E., Merlini, A. (1988c). La delimitazione di aree e regioni metropolitane: due metodi a confronto. In: Proceedings from the IX Conferenza Italiana di Scienze Regionali, AISRe, 2, Turin (pp. 677–694).

Chelli, F. M., Mattioli, E., & Merlini, A. (1992). Aspetti teorici e verifiche empiriche di alcune misure dell'urbanizzazione. In L. Di Comite & M. A. Valleri (Eds.), *Urbanizzazione e Conturbanizzazione: il caso italiano* (pp. 51–67). Bari: Cacucci Editore.

Chiucchi, M. S., Giuliani, M., & Marasca, S. (2014). The design, implementation and use of intellectual capital measurement. A case study. *Management Control, 2,* 143–168.

Chiucchi, M.S., Giuliani, M., Marasca, S. (2016). The use of intellectual capital reports: The case of Italy. *Electronic Journal of Knowledge Management, 14*(4).

Chiucchi, M.S., Giuliani, M., Marasca, S. (2018). Levers and barriers to the implementation of intellectual capital reports: A field study. *The Routledge Companion to Intellectual Capital,* Routledge, New York.

Ciommi, M., Gentili, A., Ermini, B., Gigliarano, C., Chelli, F. M., & Gallegati, M. (2017a). Have Your Cake and Eat it Too: The Well-Being of the Italians (1861–2011). *Social Indicators Research, 134*(2), 473–509.

Ciommi, M., Gigliarano, C., Emili, A., Taralli, S., & Chelli, F. M. (2017b). A new class of composite indicators for measuring well-being at the local level: An application to the equitable and sustainable well-being (BES) of the Italian provinces. *Ecological Indicators, 76,* 281–296.

Conti, G., & Modiano, P. (2012). Problemi dei paesi a sviluppo tardive in Europa: riflessioni sul caso italiano. *L'industria, XXXIII*(2), 221–235.

Crivellini, M., & Pettenati, P. (1989). Modelli locali di sviluppo. In G. Becattini (Ed.), *Modelli locali di sviluppo.* Bologna: Il Mulino.

Del Bene, L. (2000). *Criteri e strumenti di controllo gestionale nelle aziende sanitarie.* Milano: Giuffrè.

Del Bene, L. (2008). *Lineamenti di pianificazione e controllo per le amministrazioni pubbliche.* Torino: Giappichelli.

Del Bene, L. (2014). L'applicazione del d.lgs. 150/2009 negli enti locali tra opportunità e rischi. L'esperienza del comune di Livorno. *Azienda Pubblica, 27*(1), 39–56.

Del Bene, L. (2016). *Alcuni fattori di inefficacia dei sistemi di controllo nelle amministrazioni pubbliche.* Milano: Egea.

Del Bene, L., Russo, S. (2017). *The challenge of public-private partnership in healthcare.* Springer.

Erevelles, S., Fukawa, N., & Swayne, L. (2016). Big data consumer analytics and the transformation of marketing. *Journal of Business Research, 69,* 897–904.

Ermini, B., Papi, L., Scaturro, F., & Tamberi, M. (2016). Ancona: Un ateneo per dottori di ricerca in economia politica. *Economia Marche Journal of Applied Economics, XXXV,* 47–93.

Favero, C., & Papi, L. (1995). Technical efficiency and scale efficiency in the Italian banking sector: A non-parametric approach. *Applied Economics, 27,* 385–395.

Ferrell, O. C., Hair, J. F., Marshall, G. W., & Tamilia, R. D. (2015). Understanding the history of marketing education to improve classroom instruction. *Marketing Education Review, 25*(2), 159–175.

Ferrero, G. (1992). *Il marketing relazionale. L'approccio delle scuole nordiche.* Trieste: Edizioni Lint.

Fuà, G. (1980).*Problems of lagged development in OECD Europe: A study of six countries.* Document n. 2277. Paris: OECD.

Fuà, G. (1986). *Conseguenze economiche dell'evoluzione demografica.* Bologna: Il Mulino.

Fuà, G. (1991). *Orientamenti per la politica del territorio, (a cura di).* Bologna: Il Mulino.

Fuà, G., & Zacchia, C. (1983). *Industrializzazione senza fratture.* Bologna: Il Mulino.

Gigliarano, C., Balducci, F., Ciommi, M., & Chelli, F. (2014). Going regional: An index of sustainable economic welfare for Italy. *Computers, Environment and Urban Systems, 45,* 63–77.

Giuli, M., & Pascucci, F. (2014). *Il ritorno alla competitività dell'espresso italiano. Situazione attuale e prospettive future per le imprese della torrefazione di caffè.* Milano: Franco Angeli.

Giuliani, M., & Marasca, S. (2011). Construction and valuation of intellectual capital: A case study. *Journal of Intellectual Capital, 12*(3), 377–391.

Giuliani, M., & Marasca, S. (2018). *La valutazione degli intangibles aziendali. Studi di Valutazione d'Azienda.* Milano: Giuffrè Editore.

Gregori, G. L. (1995). *Aspetti economici e gestionali delle relazioni tra imprese industriali ed intermediari commerciali.* Torino: Giappichelli Editore.

Gregori, G. L. (2016). *Made in Italy. Una lettura critica fra eredi virtuosi e dissipatori.* Bologna: Il Mulino.

Gregori, G. L., & Cardinali, S. (2015). *Wellbeing marketing: profili di ricerca e nuovi strumenti di gestione nel mercato del benessere.* Milano: Franco Angeli.

Gregori, G. L., Cardinali, S., & Temperini, V. (2012). *Traiettorie di sviluppo delle imprese calzaturiere nel nuovo contesto competitivo.* Torino: Giappichelli Editore.

Gregori, G. L., Pascucci, F., & Cardinali, S. (2016). *Internazionalizzazione digitale.* Milano: Franco Angeli.

Grönroos, C. (2004). The relationship marketing process: Communication, interaction, dialogue, value. *Journal of Business & Industrial Marketing, 19*(2), 99–113.

Gruppo di Ancona. (1999). *Trasformazioni dell'economia e della società italiana.* Bologna: Il Mulino.

Gummesson, E. (1997). Relationship marketing as a paradigm shift: Some conclusions from the 30R approach. *Management Decision, 35*(4), 267–272.

Hogan, J. E., Lemon, K. M., & Rust, R. T. (2002). Customer equity management: Charting new directions for the future of marketing. *Journal of Service Research, 5*(4), 4–12.

Lamonica, G. R., & Chelli, F. M. (2018). The performance of non-survey techniques for constructing sub-territorial input–output tables. *Papers in Regional Science, 97*(4), 1169–1202.

Lamonica, G. R., & Zagaglia, B. (2013). The determinants of internal mobility in Italy, 1995–2006: A comparison of Italians and resident foreigners. *Demographic Research, 29,* 407–440.

Leeflang, P. S. H., Verhoef, P. C., Dahlström, P., & Freundt, T. (2014). Challenges and solutions for marketing in a digital era. *European Management Journal, 32,* 1–12.

Liu, X., & Burns, A. C. (2018). Designing a marketing analytics course for the digital age. *Marketing Education Review, 28*(1), 28–40.

Lucchetti, R., Papi, L., & Zazzaro, A. (2001). Banks' inefficiency and economic growth: A micro-macro approach. *Scottish Journal of Political Economy, 48,* 400–425.

Mandelli, A. (2017). *Big data marketing. Creare valore nella platform economy con dati, intelligenza artificiale e IoT.* Milano: Egea.

Marasca, S. (1989). *Il controllo di gestione nelle aziende commerciali complesse.* Torino: Giappichelli.

Marasca, S. (1991). Le determinazioni dei costi di prodotto: alcune considerazioni sulle metodologie di calcolo. *Rivista Italiana di Ragioneria e di Economia Aziendale, 1*(2), 35–45.

Marasca, S. (1993). *Fabbisogno conoscitivo e determinazioni di costo nei processi decisionali delle piccole e medie imprese.* Ancona: Clua.

Marasca, S. (1998). *L'introduzione del controllo di gestione: profili teorici e problemi applicativi.* Rimini: Maggioli.

Marasca, S. (1999). *Le valutazioni nel bilancio d'esercizio.* Torino: Giappichelli.

Marasca, S. (2001). *Sistemi di controllo e sistemi di valutazione: criticità e prospettive nelle Amministrazioni Pubbliche Locali.* Torino: Giappichelli.

Marasca, S. (2003). I sistemi di contabilità analitica nelle indagini empiriche: una lettura diacronica. *Contabilità e Cultura Aziendale, 3*(1), 13–35.

Marasca, S. (2016). Achieving the efficiency of plasmapheresis. The contribution of cost accounting and cost management. *Blood Transfusion, 14,* S88–S93.

Marasca, S., Comuzzi, E., & Olivotto, L. (2009). *Intangibles. Profili di gestione e misurazione.* Milano: Franco Angeli.

Marasca, S., D'Andrea, A., & Piani, M. (2013). I costi congiunti degli emocomponenti: il caso della Regione Marche. *Mecosan, 22*(88), 61–74.

Marasca, S., Eandi, M., Povero, M., D'Andrea, A., Tieghi, A., Randi, V., et al. (2016). Efficiency of plasmapheresis: A comparison of three Italian centers. *Blood Transfusion, 14*(2), 83–87.

Marchi, L., & Marasca, S. (1994). *Il bilancio civilistico-fiscale: i nuovi principi di classificazione e valutazione.* Milano: EBC.

Marchi, L., & Marasca, S. (2010). *Le risorse immateriali nell'economia delle aziende.* Bologna: Il Mulino.

Marchi, L., Marasca, S., & Riccaboni, A. (2009). *Controllo di gestione. Metodologie e strumenti.* Arezzo: Knowità Editore.

Marchi, L., Zavani, M., & Branciari, S. (2010). *Economia dei gruppi e bilancio consolidato. Una interpretazione degli andamenti economici e finanziari.* Torino: Giappichelli.

Marcone, M. R. (2004). *La competitività delle PMI italiane nella subfornitura internazionale. Il caso delle imprese produttrici di circuiti stampati.* Torino: Giappichelli Editore.

Marcone, M. R. (2013). Styling and design in multi-segmented market strategies: The case of the italian knitwear sector. *International Journal of Learning and Change, 7*(1/2), 86–103.

Mariani, F., Recchioni, M. C., & Ciommi, M. (2018). Merton's portfolio problem including market frictions: A closed-form formula supporting the shadow price approach. *European Journal of Operational Research, 275*(3), 1178–1189. https://doi.org/10.1016/j.ejor.2018.12.022.

Mattioli, E., & Lamonica, G.R. (2010). On the similarity of methods for ranking alternatives and making transitive the index numbers. *Italian Journal of Applied Statistics, 22*(2).

Mattioli, E., Lamonica, G.R. (2012). Degree of integration and activation power of tourism in the european countries' economies: An input output analysis. *Rivista Italiana di Economia Demografia e Statistica, 66*(2).

Mattioli, E., & Lamonica, G. R. (2015a). The world's economic geography: Evidence from the world input–output table. *Empirical Economics, 50*(3), 697–728.

Mattioli, E., & Lamonica, G. R. (2015b). The impact of the tourism industry on the world's largest economies: An input–output analysis. *Tourism Economics, 21*(2), 419–426.

Mattioli, E., & Lamonica, G.R. (2015c). The evolution of the vertical specialization in the world economy (1995–2011). *Rivista Italiana di Economia, Demografia e Statistica, 69*(3).

Mattioli, E., & Merlini, A. (1983). Alcuni aspetti metodologici nello studio delle migrazioni interne. L'esperienza italiana 1969–1979, Statistica, anno XLIII, (2).

Mayer-Schönberg, V., & Cukier, K. (2013). *Big data: A revolution that will transform how we live, work, and think.* New York: Houghton Mifflin Harcourt.

Merlini, A. (1978). Sui procedimenti di discriminazione parametrici basati su stimatori corretti in media e consistenti delle funzioni di densità normali. *Rivista Italiana di Economia Demografia e Statistica,* 3–4.

Möller, K. (2013). Theory map of business marketing: Relationships and network perspectives. *Industrial Marketing Management, 42,* 324–335.

Monticelli, C., & Papi, L. (1996). *European integration, monetary co-ordination and the demand for money.* Oxford: Claredon Press.

Niccoli, A. (1989). *Credito e sviluppo.* Milano: Giuffrè editore.

Niccoli, A., & Papi, L. (1993). *Debito pubblico e sistema finanziario.* Milano: Giuffrè editore.

Pascucci, F. (2013). *Strategie di marketing online per il vantaggio competitivo aziendale.* Bologna: Esculapio Editore.

Pascucci, F. (2018). The export competitiveness of Italian coffee roasting industry. *British Food Journal, 120*(7), 1529–1546.

Pascucci, F., Cardinali, S., Gigliarano, C., & Gregori, G. L. (2017). Internet adoption and usage: Evidence from Italian micro enterprises. *International Journal of Entrepreneurship and Small Business, 30*(2), 259–280.

Pascucci, F., & Temperini, V. (2017). *Trasformazione digitale e sviluppo delle PMI. Approcci strategici e strumenti operativi.* Torino: Giappichelli Editore.

Perna, A., Cardinali, S., & Gregori, G. L. (2013). Coping with alternatives in sales organisations: Experiences from an Italian company. *Journal of Business Market Management, 6*(3), 107–122.

Perna, A., Runfola, A., Guercini, S., & Gregori, G. L. (2015). Relationship beginning and serendipity: insights from an Italian case study. *The IMP Journal, IX*(3), 233–249.

Recchioni, M. C., & Sun, Y. (2016). An explicitly solvable Heston model with stochastic interest rate. *European Journal of Operational Research, 249,* 359–377.

Recchioni, M. C., Sun, Y., & Tedeschi, G. (2017). Can negative interest rates really affect option pricing? Empirical evidence from an Explicitly Solvable Stochastic Volatility Model. *Quantitative Finance, 17*(8), 1257–1275.

Recchioni, M. C., & Tedeschi, G. (2017). From bond yield to macroeconomic instability: A parsimonious affine model. *European Journal of Operational Research, 262,* 1116–1135.

Runfola, A., Perna, A., Baraldi, E., & Gregori, G. L. (2017). The use of qualitative case studies in top business and management journals: A quantitative analysis of recent patterns. *European Management Journal, 35*(1), 116–127.

Schwab, K. (2015). *The fourth industrial revolution.* New York: World Economic Forum.

Silvestrelli, S. (1979). *Lo sviluppo industriale delle imprese produttrici di mobili in Italia.* Milano: Franco Angeli.

Silvestrelli, S., & Bellagamba, A. (2017). *Fattori di competitività dell'impresa industriale. Un'analisi economica e manageriale.* Torino: Giappichelli Editore.

Sterlacchini, A. (2008). R&D, higher education and regional growth: Uneven linkages among European regions. *Research Policy, 37*(6–7), 1096–1107.

Tedeschi, G., Recchioni, M. C., & Berardi, S. (2018). An approach to identifying micro behavior: How banks' strategies influence financial cycles. *Journal of Economic Behavior and Organization, 162,* 329–346. https://doi.org/10.1016/j.jebo.2018.12.022.

Vargo, S. L., & Lusch, R. F. (2004). Evolving to a new dominant logic for marketing. *Journal of Marketing, 68*(1), 1–17.

Vargo, S. L., & Lusch, R. F. (2008). Service-dominant logic: Continuing the evolution. *Journal of the Academy of Marketing Science, 36*(1), 1–10.

Vitali, O. (1983). *L'evoluzione urbano-rurale in Italia 1951–1977.* Milano: F. Angeli.

Youssef, Y., Johnston, W. J., Abdelhamid, T. A., Dakrory, M. I., & Seddick, M. (2018). A customer engagement framework for a B2B context. *Journal of Business & Industrial Marketing, 33*(1), 145–152.

The School of Agent-Based Models in Ancona

Mauro Gallegati, Antonio Palestrini and Alberto Russo

Abstract This chapter follows the path taken by our research group to extend economic modeling to the analysis of complex systems. It began with research meetings in the old Economic Department of the Faculty of Economics in the 1990s, where we imagined possible ways of introducing heterogeneity in traditional economic analysis arriving to the present day with the latest research which analyses interaction networks between economic agents (e.g., bank-firm), underlining the importance of composition effects and the emergence of macro-level properties that cannot be limited to the behavior of a "representative agent". In fact, from the very beginning this research group has believed in the importance of analyzing the distribution of the characteristics of economic individuals (income, wealth, etc.) and of their direct interactions mediated by a network. Recent developments, using the agent-based methodology (i.e. an approach with interacting heterogeneous agents that simulates, through the use of computers, the evolution of complex economic systems), focus on the role of financial fragility, inequality, systemic risk and the crises of the capitalist system, extending the framework to an increasingly broader macroeconomic model. One of the main results concerns the role of interaction between financial factors and the real economy, through the analysis of the effects of network between the different micro-entities of the system and the potential contagion effect through the financial economics channels, also highlighting the positive role of regulation. The research group has also shown the relevance of the combination of financial fragility and inequality, two interrelated phenomena that have characterized the economic development of Western countries in recent decades and are at the root of understanding the Great Recession. Another line of research concerned the application of mathematical-statistical tools, used in particular in physics, for the analysis of economic and financial phenomena, contributing to the development

M. Gallegati · A. Palestrini (✉) · A. Russo
DIMA-UNIVPM, Piazzale Martelli 8, Ancona, Italy
e-mail: a.palestrini@univpm.it

M. Gallegati
e-mail: mauro.gallegati@univpm.it

A. Russo
e-mail: alberto.russo@univpm.it

© Springer Nature Switzerland AG 2019
S. Longhi et al. (eds.), *The First Outstanding 50 Years
of "Università Politecnica delle Marche"*,
https://doi.org/10.1007/978-3-030-33879-4_2

23

of Econophysics. This research group aims to continue, over the next few years, to provide new ideas and tools to study the macroeconomic dynamics of heterogeneous agents that must coordinate in a complex system.

1   The Origins: Financial Fragility and Power Law Distributions

In the second half of the 90s a small group of researchers and Ph.D. students in the Department of Economics of the then University of Ancona met on the fifth floor of Palazzo degli Anziani (the old University building situated below the Church of San Ciriaco) to discuss how to change the macroeconomic approach that was emerging in those years. This approach, undoubtedly proliferative, was based on the hypothesis of the representative agent, perfect markets, absence of direct interactions and coordination problems in non-equilibrium dynamics. Although understanding the need for simplification, our belief was, in fact, that heterogeneity with its problems of aggregation, direct interaction and adjustments to equilibrium needed to be addressed with greater attention.[1] Our intuition, later confirmed by subsequent analyses, is that often aggregated phenomena emerge spontaneously from the interactions of individuals struggling to coordinate their actions on markets: macroscopic regularities emerge from microscopic behavior. In other words, aggregate "laws" are due to emergence rather than to microscopic rules. In turn, emergent macroeconomic dynamics feeds back on microeconomic behavior through a downward causation process, in which economic and social structures affect the evolution of opportunities and preferences characterizing microeconomic units. This research group has grown from the long-lasting interaction of researchers with different skills and sensibilities thanks also to the organization of the WEHIA conference (Workshop on Economies with Heterogeneous Interacting Agents), which started at the University of Ancona in 1996. The success was beyond expectations. More than twenty years ago, we certainly did not expect that so many researchers would share our own perplexities and ideas of innovation. At present, WEHIA is a well-established yearly international forum for discussion and cross-fertilization of ideas on social interaction of heterogeneous agents.

A collaboration between the founder of this group Mauro Gallegati with Antonio Palestrini, and Domenico Delli Gatti from the Catholic University in Milan was decisive in setting the reference model to analyse heterogeneous agent behaviors. The model was formalized in Delli Gatti et al. (2000), a work in which business cycles are affected by the evolution over time of the distribution of heterogeneous agents, classified by the degree of financial fragility. Not only is the distribution of agents important to understand aggregate phenomena, but it is also an outcome of interacting heterogeneous agents. In the work Delli Gatti et al. (2003) we were able to introduce in the analysis the entry-exit industrial dynamics in order to investigate

---

[1] See Leijonhufvud (2006) and the discussion in De Vroey (2016).

how it affects the dynamics of the distribution, and the macroeconomic performance, of firms which are differentiated by the equity ratio, i.e. the ratio of net worth to the capital stock. These studies use theoretical results of the '90s of Bruce Greenwald and the Nobel Price winner Joseph E. Stiglitz (Greenwald and Stiglitz 1993) and the numerous discussions we had with them at the Columbia University and around the world. Such discussions were very important for the analysis in the highly cited paper by Delli Gatti et al. (2005) in which, for the first time, we showed that a simple financial fragility agent-based model, based on complex interactions of heterogeneous agents, is able to replicate a large number of scaling type stylized facts with a remarkably high degree of statistical precision. The statistical properties of such kind of distributions were subsequently analyzed in Palestrini (2007), Delli Gatti et al. (2008b), and Gallegati and Palestrini (2010).

Our analysis also explored the functioning of financial markets populated by heterogeneous agents using genetic algorithms and reinforcement rules. The findings in Chiarella et al. (2003) and Bischi et al. (2006) show situations of multistability generated by herding behaviors. The instability of financial markets is also characterized by strong path dependence.

In a subsequent work (Gallegati et al. 2011) we were able to show that the ABM, an approach applied to financial markets, is also able to explain a famous stylized fact described by Kindleberger (2000): The period of financial distress following the peak and preceding the crash of a bubble.

Returning to the analysis of the link between the real and the financial part of an economic system, during the second part of the 2000s the collaboration between UNIVPM (Mauro Gallegati and Alberto Russo) and the University of Cattolica (Domenico Delli Gatti) researchers, with Greenwald and Stiglitz at the Columbia University, focuses on the idea that business cycle fluctuations are strictly related to the network structure of heterogeneous-agent economies. A first step in this direction was a model with a *static* network of bank-bank (interbank market), bank-firm (bank credit) and firm-firm (trade credit) interrelations through productive and credit relationships (Delli Gatti et al. 2006). An extension with an *evolving* network structure, aimed at studying financially constrained business fluctuations, was published in an NBER working paper (Delli Gatti et al. 2008a), and eventually led to a paper which opened a new stream of research and received a lot of citations in recent years (Delli Gatti et al. 2010b). This latter work describes an economy composed of heterogeneous firms, belonging to two sectors (upstream and downstream firms), which interact with financially fragile heterogeneous banks in order to finance their production, giving rise to a *network-based financial accelerator*. Simulation results feature financial contagion dynamics and bankruptcy avalanches, stressing the fundamental role of financial factors and their impact on production and employment dynamics. Along these lines, the interplay between firms' leverage and the business cycle has been further investigated in Riccetti et al. (2013a), as well as stock market dynamics in Riccetti et al. (2016b). Furthermore, this modelling framework aimed at studying business cycles within evolving credit networks has been validated with empirical micro-data on firm-bank relationships, firstly in Bargigli et al. (2014), through a network calibration of firm-bank interconnections, and then in Bargigli et al. (2018), by proposing a meta-modelling approach.

2 Toward a Fully-Fledge Agent-Based Stock-Flow Consistent Macroeconomic Model

In the meanwhile, a first step toward the development of a fully-fledged agent-based macroeconomic model has been proposed by modelling the interaction among heterogeneous agents through a decentralized matching protocol, with the aim of investigating the potential role of fiscal policy on innovation dynamics and long-term economic growth (Russo et al. 2007). In fact, while Delli Gatti et al. (2010b) proposed a *partial* disequilibrium approach to study financially constrained business cycle fluctuations, this new approach moved toward a *general* disequilibrium economic model, by introducing all the typical ingredients of a macroeconomic framework. This kind of approach was first proposed in Riccetti et al. (2015) by applying a *decentralized matching mechanism* to all markets that compose the economic system. Basically, each agent on the demand side looks for a potential partner on the supply side (within a subset of the population, due to imperfect information), trying to find out the best match (namely the lowest price) for the demanded goods or services. In this paper the interplay between finance and the real economy is at the core of endogenous business cycles: the expansion of finance is beneficial for macroeconomic performance up to a certain point; then, when agents' leverage becomes "excessive," the relation between finance and the real economy tends to be inverted, due to the increase of non-performing loans and the possible default of both firms and banks, so that the negative effects of financial contagion result in a reduction of incomes and an increase in unemployment. Moreover, the model is able to endogenously generate *extended crises* due to a lack of aggregate demand in a Keynesian fashion. The complex dynamics involved in extended crises was then investigated by Delli Gatti et al. (2012) by stressing the role of sectoral dislocation. In particular, this paper points out that productivity growth in the "traditional sector" (agriculture during the Great Depression, manufacturing during last decades) may result in a large crisis, in case of *mobility* constraints which limit the transition toward the "advanced sector" (manufacturing then, services recently). Further extensions of the benchmark agent-based macro model with decentralized matching (Riccetti et al. 2015) have been proposed for investigating various topics such as the stabilization role of unemployment benefits (Riccetti et al. 2013b), the financialization of non-financial firms (Riccetti et al. 2016a), inequality and consumer credit (Russo et al. 2016), financial regulation (Riccetti et al. 2018), and the effectiveness of monetary policy in "double-dip recession" episodes (Giri et al. 2019).

In late 2013 a joint research initiative was launched with the research group of Prof. Stephen Kinsella at the Kemmy Business School in Limerick (Ireland), and Prof. Stiglitz (Columbia University, New York, USA) to realize a merge between the Agent Based approach to macroeconomic modeling (Delli Gatti et al. 2011) and the accounting-based modeling logic of the Post-Keynesian Stock Flow Consistent (SFC, hereafter) framework (Godley and Lavoie 2007). This AB-SFC research program was funded by the Institute for New Economic Thinking (INET), being part of Prof. Joseph Stiglitz's Task Force on Macroeconomic Efficiency and Instability at INET. The task force aimed at developing alternative tools for macroeconomic analysis after the financial turmoil of 2007 and the ensuing Great Recession had raised serious doubts on the plausibility and efficacy of standard ones, in particular the class of DSGE models. By bridging the AB and the SFC approaches the project aimed to model current economic systems as complex adaptive systems, while at the same time providing a comprehensive and fully integrated representation of the real and financial sides of an economic system, and their mutual relationships. This collaboration led to the development of a first 'benchmark' AB-SFC macro, presented in Caiani et al. (2016). Besides presenting and validating the model, the paper aimed at sketching out the main traits and 'rules to follow' of the novel AB-SFC modeling paradigm. The model and the methodology proposed in the paper provided the basis for several other developments: in Caiani et al. (2019a) an augmented version of the 'benchmark' with endogenous technological change and workers' hierarchical differentiation was presented to analyze the relationship between income and wealth inequality on the one hand, and the macroeconomic performance on the other hand. Caiani et al. (2019b) employs the Kriging meta-modeling technique (originally developed in Geo-physics) to perform an extensive analysis of the impact of alternative wage growth regimes in presence of different characterizations of firms' investment behavior and different characterizations of the technological change process. Schasfoort et al. (2018) provides a further extension of the 'benchmark' model encompassing an interbank market in order to analyze the importance of different monetary policy transmission channels. Finally, another augmented version of the 'benchmark' to study the effects of the diffusion of automation and robots in manufacturing productive processes, and their possible impact on employment, job polarization, and growth, is currently under development. Besides the continuous refining of the benchmark model, the AB-SFC research line has also resulted in the development of several other brand new models: Caiani et al. (2018a) for example provides a fully fledged multi-country AB-SFC model aimed at describing the European Monetary Union (EMU). The model was employed to analyze the effects of fiscal discipline on the economic performance and resilience of a Monetary Union comparable, in its institutional architecture, to the EMU. Using a slightly amended version of this model Caiani et al. (2018b) accounts for the twofold role of salaries paid to workers, which represents at the same time a fundamental source of aggregate demand and a determinant of firms' international cost competitiveness, in order to investigate how alternative wage growth patterns impact on the economic dynamics of an artificial Monetary Union. In the future, two further refinements of the model are scheduled and dedicated, respectively, to improve the description of the financial side of the model so to provide a better account of the fundamental role played by international financial flows, and to provide a more realistic assessment of the determinants of international trade by accounting for countries' structural specialization and product complexity.

Another strand of research applies the AB-SFC approach to the analysis of the impacts of financial dynamics and financial innovations on the economic system. Botta et al. (2019) presents a hybrid AB-SFC macroeconomic model, with all sectors at a macro-aggregated level, with the exception of the households sector. This modelling choice is motivated by two considerations. On the one hand, the parsimonious use of micro characterization allows to obtain a much clearer interpretation of simulation results. On the other hand, the assumption of multiple heterogeneous households is necessary for the purpose of the model and in particular for the analysis of financial contagion and inequality, in the context of a developed financial system. Indeed, the model—building on Botta et al. (2015) and Botta et al. (2018)—focuses on the system securitization and on its impact on wealth and income inequality. Simulation results suggest that securitization may determine an increase in economic growth, favoured by a higher level of credit supply. This however, may come at the cost of a more unequal and financially unstable economic system. The model aims at representing the first step in a promising research path. It will soon be adapted to the study of different financial phenomena such as the Repo market, the safe asset gap, and the housing market.

3   Sideways: Networks, Financial Crises, and Experimental Economics

Our research group also analyzes the systemic risk problem and the macro-prudential policies aiming at reducing economic system vulnerability that has been at the center of economic debate of the last few years (Commitee 2011; Yellen 2011; Angelini et al. 2012). Credit networks play a crucial role in diffusing and amplifying local shocks, thus we try to define both early warning indicators of crises and policy precautionary measures based on the analysis of the dynamics of credit network connectivity. Following the network-based financial accelerator approach (Delli Gatti et al. 2005, 2010a; Battiston et al. 2012).

In Catullo et al. (2015) we constructed an agent based model reproducing an artificial credit network populated by heterogeneous firms and banks. The model is calibrated on a sample of firms and banks quoted on the Japanese stock-exchange markets from 1980 to 2012. The simulated credit network evolves reproducing endogenously the credit cycle dynamics and the occurrence of crises. This framework allows us to isolate an early warning measure for crises. This measure is defined as the level of concentration of bank connectivity and leverage. Indeed, if few banks are strongly connected to the others and, at the same time, they show a relatively high level of leverage, even a small local shock that hits one of these more connected banks may trigger a crisis.

In Catullo et al. (2017), following the methodology developed by Schularick and Taylor (2012), we found that both credit and connectivity growth rates are positively correlated with crisis probability and that they are effective early warning measures in both empirical and simulated data. Moreover, the model is suitable for designing macro prudential policies which exploit agent heterogeneity and the networks interaction structure. Indeed, capital related measures which force banks to avoid lending to more indebted firms may decrease output volatility without causing consistent credit reductions and, thus, output contractions. We also tested permanent capital-related measures applied to larger and more connected banks only. When interventions target banks that are relatively central in the credit network in terms of size and connections, the vulnerability of the economic system may be substantially reduced without affecting aggregate credit supply and output.

Thus, the analysis of credit network connectivity may be useful for assessing the emergence of system risk. Besides, agent-based models that endogenize credit network dynamics may be used for testing the effectiveness of early warning indicators and the effects of different macro-prudential policies.

A promising line of research is the one exploiting the links between experimental economics and ABM. Laboratory experiments and ABM are instruments to investigate the behaviour of heterogeneous agents operating in a dynamical environment. Even if these tools share a common ground, they are usually seen as independent fields. The first attempt to merge those disciplines was the work by Arthur (1991). This essay marked the starting point of a new stream of literature whose aim is to implement experimental data to calibrate/validate ABMs. Marimon et al. (1993) proposed a new experimental design useful to directly observe expectations: the Learning to Forecast Experiment. Subjects in this experiment play the role of professional forecasters so that they only submit predictions. Such a basic framework allows to elicit expectations without the noise due to trading activity. This turned out to be a powerful tool to analyze expectations in financial markets (Hommes et al. 2005) as well as in commodity markets (Bao et al. 2013) and macroeconomic framework (Assenza et al. 2011). The data gathered from these experiments were implemented to build up a simple model of heterogeneous agents named Heuristic Switching Model. The underlying idea of this model is that markets are formed by agents with heterogeneous expectations and the market price is simply the result of their interaction (Anufriev et al. 2013). To validate the theoretical findings presented in Palestrini and Gallegati (2015), we ran, in 2014, a Learning to Forecast Experiment in the lab of our faculty. Results were very useful to understand the process of expectation formation in financial markets with non-constant fundamental value. Indeed, we were able to confirm in Colasante et al. (2017) the hypothesis put forward by Palestrini and Gallegati (2015) that expectations are mainly represented by modified adaptive rule and, moreover, subjects are able to learn a trend in the fundamental price. The follow up of this initial project culminated in the design of a new experimental setting jointly with the research group based in Castellon (Spain). This represents the first experiment capable of eliciting both short- and long-term expectations (Colasante et al. (2018a)). After replicating the heuristic switching model, elicited long-term predictions were implemented to calibrate a model of heterogeneous agents capable of explaining the regularities observed in the laboratory (Colasante et al. 2018c). Agent predictions seem to be better explained by a non-parametric approach rather than a parametric one such as the hsm. The main finding, in fact, is that the last realized price in financial markets plays a pivotal role in determining short as well as long-term expectations. Finally, the potential of this experimental design has been implemented to study expectations in commodity markets (Colasante et al. 2018b). Besides corroborating the main finding observed in another experiment (Heemeijer et al. 2009), the analysis in terms of both convergence to the fundamental value and coordination of expectations pave the way to a different concept of heterogeneity. By having the opportunity to observe also long-term expectations, we stressed the importance of measuring the forecast disagreement not only between different subjects but foremost across forecasting horizons.

Finally, the ABM approach allows us to deeply investigate in the last decades profound structural change, which caused by the transaction from a manufacturing economy to a service-based one, is one of the causes of the current crisis (see Delli Gatti et al. 2012). The privatization of the banking system with the consequent redirection of banking activity from the credit sector to the financial one, the liberalization of financial markets, the globalization and the delocalisation of production with the resulting flexibility in the labor market are just some of the many transformations affecting the socio-economic system in recent decades. All these profound changes have been poorly described or simply ignored by the "static" mainstream economics. Based on the lessons taken from the recent socio-economic transformations, Gallegati with Grilli and Tedeschi have analyzed "the missing link" between the micro and the macro dynamics. Therefore, what their scientific research proposes is the formalization of a descriptive and non-normative theory, able to grasp the complex dynamics of social systems. The authors restore in the analysis fundamental features of the economic cycles based on the effects of shocks on the network of credit relationships which may lead to systemic risk, i.e. the risk of an epidemic diffusion of financial distress and eventually to a fully fledged financial crisis (Grilli et al. 2014, 2015a, b). First and foremost, in the aggregate view (where no interaction takes place) by construction the shock which originates the fluctuation is aggregate i.e. uniform across agents. Therefore big crises are necessarily originated by big shocks. In the real world, however, an idiosyncratic shock can well be the source of an epidemic diffusion of financial distress. In other words, in a financial network idiosyncratic shocks usually do not cancel out in the aggregate, especially if they hit crucial nodes (hubs) of the network. Secondly, in a credit network the interaction can lead to an avalanche of bankruptcies (Tedeschi et al. 2012). Suppose, for instance, that a firm goes bust. Both the suppliers and the banks which made business with the bankrupt firm will bear the brunt of the default. In other words, the default creates a negative externality for connected agents. The deterioration of the bank's financial condition due to the borrower's bankruptcy may be absorbed if the size of the loan is small and/or the bank's net worth is high. If this is not the case, also the bank goes bankrupt. If the bank survives, however, it will restrain credit supply and/or make credit conditions harsher—raising the interest rate on loans across the board—for all its borrowers. Therefore, the default of one agent can bring about an avalanche of bankruptcies. While the proximate cause of the bankruptcy of a certain firm in the middle of the avalanche is the interest rate hike, the remote cause is the bankruptcy of a firm at the beginning of the avalanche that forced the banks to push interest rates up. The interest rate hike leads to more bankruptcies and eventually to a bankruptcy chain: "the high rate of bankruptcy is a cause of the high interest rate as much as a consequence of it" (Greenwald and Stiglitz 2003, p. 145). An avalanche of bankruptcies therefore is due to the positive feedback of the bankruptcy of a single agent on the net worth of the "neighbors", linked to the bankrupt agent by credit links of one sort or another. Finally, asset price changes in mainstream models (i.e. New Keynesian or Dynamic Stochastic General Equilibrium models) with financial factors are almost invariable bounded by a built in mean reverting tendency. It is indeed very difficult to account for asset price booms and busts in these models because of the restrictive way expectations are modelled. Rational expectations and maximizing rational agents cannot account for bubbles and their repercussions on financial markets and systemic risk. These are, however, the crucial aspects of interactions and systemic risk addressed by Tedeschi et al. (2009, 2012) and Recchioni et al. (2015) in their financial price models. Therefore, vision of the three authors tries to go beyond the current state of the art in economics and focuses on the complex pattern of relationships which is a natural research issue to be dealt with by means of network analysis.

Summarizing this short survey, more than half a century ago, Orcutt (1957) proposed *microsimulation* as a methodology to study economic systems, trying to reproduce both micro and macro patterns. Microsimulation is a methodology which is used in various scientific fields to investigate the states and behaviors of heterogeneous units, like individuals, firms, households, which interact and evolve in a certain environment. The exponential growth of computational power has allowed research groups, like the one at UNIVPM, to develop complex macro models. In the future, more and more computational power will be available and this will allow to further extend the use of agent-based models to replicate, with increasing accuracy, the real world. Big data will be used to match simulation findings and empirical evidence, and the development of econometrics and validation methods will allow us to better forecast the future. Experimental economics and neuroeconomics will certainly play a role in suggesting a deeper understanding of economic agents' choices, the role of emotions and the complicated interactions of neural processes at the basis of both individual and collective actions. In our view, Agent-Based Modelling will continue to be a useful methodology for investigating the evolution of complex adaptive systems like the economy.

Meanwhile the group was able to finance 21 postdocs; obtain 12 EU projects; 3 INET and 4 MIUR grants. We organized 12 International Workshops and 3 ABM summer schools. Great merit of these achievements is due to the group of young researchers working in our Faculty of Economics: Eugenio Caverzasi, Alessandro Caiani, Ermanno Catullo, Fabio Clementi, Annarita Colasante, Corrado Di Guilmi, Lisa Gianmoena, Federico Giri, Ruggero Grilli, Luca Riccetti, and Gabriele Tedeschi.

Without their help handling the complexity of these economic models would have been overwhelming.

References

Angelini, P., Nicoletti-Altimari, S., & Visco, I. (2012). *Macroprudential, microprudential and monetary policies: Conflicts, complementarities and trade-offs.* Questioni di Economia e Finanza (Occasional Papers) 140, Bank of Italy, Economic Research and International Relations Area.

Anufriev, M., Hommes, C. H., & Philipse, R. H. (2013). Evolutionary selection of expectations in positive and negative feedback markets. *Journal of Evolutionary Economics, 23*(3), 663–688.

Arthur, W. B. (1991). Designing economic agents that act like human agents: A behavioral approach to bounded rationality. *The American Economic Review, 81*(2), 353–359.

Assenza, T., Heemeijer, P., Hommes, C., & Massaro, D. (2011). *Individual expectations and aggregate macro behavior.* Dnb working papers, Netherlands Central Bank, Research Department.

Bao, T., Duffy, J., & Hommes, C. (2013). Learning, forecasting and optimizing: An experimental study. *European Economic Review, 61*, 186–204.

Bargigli, L., Gallegati, M., Riccetti, L., & Russo, A. (2014). Network analysis and calibration of the leveraged network-based financial accelerator. *Journal of Economic Behavior & Organization, 99*(C), 109–125. https://doi.org/10.1016/j.jebo.2013.12.01. https://ideas.repec.org/a/eee/jeborg/v99y2014icp109-125.html.

Bargigli, L., Gallegati, M., Riccetti, L., & Russo, A. (2018). Network calibration and metamodeling of a financial accelerator agent based model. *Journal of Economic Interaction and Coordination.* https://doi.org/10.1007/s11403-018-0217-8.

Battiston, S., Delli Gatti, D., Gallegati, M., Greenwald, B. C., & Stiglitz, J. E. (2012). Liaisons dangereuses: Increasing connectivity, risk sharing, and systemic risk. *Journal of Economic Dynamics & Control, 36*, 1121–1144.

Bischi, G. I., Gallegati, M., Gardini, L., Leombruni, R., & Palestrini, A. (2006). Herd behavior and nonfundamental asset price fluctuations in financial markets. *Macroeconomic Dynamics, 10*(4), 502–528.

Botta, A., Caverzasi, E., & Tori, D. (2015). Financial-real-side interactions in an extended monetary circuit with shadow banking: Loving or dangerous hugs? *International Journal of Political Economy, 44*(3), 196–227.

Botta, A., Caverzasi, E., Russo, A., Gallegati, M., & Stiglitz, J. E. (2019). Inequality and finance in a rent economy. *Journal of Economic Behavior and Organization.* https://doi.org/10.1016/j.jebo.2019.02.013.

Botta, A., Caverzasi, E., & Tori, D. (2018). The macroeconomics of shadow banking. *Macroeconomic Dynamics.* https://doi.org/10.1017/S136510051800041X:1-30.

Caiani, A., Godin, A., Caverzasi, E., Gallegati, M., Kinsella, S., & Stiglitz, J. E. (2016). Agent based-stock flow consistent macroeconomics: Towards a benchmark model. *Journal of Economic Dynamics & Control, 69*, 375–408.

Caiani, A., Catullo, E., & Gallegati, M. (2018a). The effects of fiscal targets in a monetary union: A multi-country agent based-stock flow consistent model. *Industrial and Corporate Change Available Online, Printed Version Forthcoming.* https://doi.org/10.1093/icc/dty016.

Caiani, A., Catullo, E., & Gallegati, M. (2018b). *The effects of alternative wage regimes in a monetary union: A multi-country agent based-stock flow consistent model.* WP Available on SSRN.

Caiani, A., Russo, A., & Gallegati, M. (2019a). Does inequality Hamper innovation and growth? *Journal of Evolutionary Economics* (forthcoming), 39.

Caiani, A., Russo, A., & Gallegati, M. (2019b). Are higher wages good for business? An assessment under alternative innovation and investment scenarios. In *Macroeconomic Dynamics.* Available online, printed version forthcoming, 40. https://doi.org/10.1017/S1365100518000299.

Catullo, E., Gallegati, M., & Palestrini, A. (2015). Towards a credit network based early warning indicator for crises. *Journal of Economic Dynamics and Control, 50*(C), 78–97.

Catullo, E., Palestrini, A., Grilli, R., & Gallegati, M. (2017). Early warning indicators and macroprudential policies: A credit network agent based model. *Journal of Economic Interaction and Coordination.*

Chiarella, C., Gallegati, M., Leombruni, R., & Palestrini, A. (2003). Asset price dynamics among heterogeneous interacting agents. *Computational Economics, 22*(2–3), 213–223.

Colasante, A., Palestrini, A., Russo, A., & Gallegati, M. (2017). Adaptive expectations versus rational expectations: Evidence from the lab. *International Journal of Forecasting, 33*(4), 988–1006.

Colasante, A., Alfarano, S., Camacho, E., & Gallegati, M. (2018a). Long-run expectations in a learning-to-forecast experiment. *Applied Economics Letters, 25*(10), 681–687.

Colasante, A., Alfarano, S., & Camacho-Cuena, E. (2018b). *The term structure of cross-sectional dispersion of expectations in a learning-to-forecast experiment*. Tech. rep., MPRA Working Papers n. 84835.

Colasante, A., Alfarano, S., Camacho Cuena, E., & Gallegati, M. (2018c). Long-run expectations in a learning-to-forecast-experiment: A simulation approach. *Journal of Evolutionary Economics*. https://doi.org/10.1007/s00191-018-0585-1.

Commitee, B. (2011). *Basel iii: A global regulatory framework for more resilient banks and banking systems*. Basel Committee on Banking Supervision: Tech. rep. O.

De Vroey, M. (2016). *A History of Macroeconomics from Keynes to Lucas and Beyond*. Cambridge University Press.

Delli Gatti, D., Gallegati, M., & Palestrini, A. (2000). Agents heterogeneity, aggregation, and economic fluctuations. In: *Interaction and Market Structure* (pp. 133–149). Heidelberg: Springer.

Delli Gatti, D., Gallegati, M., Giulioni, G., & Palestrini, A. (2003). Financial fragility, patterns of firms entry and exit and aggregate dynamics. *Journal of Economic Behavior & Organization, 51*(1), 79–97.

Delli Gatti, D., Di Guilmi, C., Gaffeo, E., Giulioni, G., Gallegati, M., & Palestrini, A. (2005). A new approach to business fluctuations: Heterogeneous interacting agents, scaling laws and financial fragility. *Journal of Economic Behavior & Organization, 56*, 489–512.

Delli Gatti, D., Gallegati, M., Greenwald, B., Russo, A., & Stiglitz, J. E. (2006). Business fluctuations in a credit-network economy. *Physica A: Statistical Mechanics and its Applications, 370*(1), 68–74. https://doi.org/10.1016/j.physa.2006.04.0. https://ideas.repec.org/a/eee/phsmap/v370y2006i1p68-74.html.

Delli Gatti, D., Gallegati, M., Greenwald, B. C., Russo, A., & Stiglitz, J. E. (2008a). Financially constrained fluctuations in an evolving network economy. *NBER Working Papers 14112*, National Bureau of Economic Research, Inc. https://ideas.repec.org/p/nbr/nberwo/14112.html.

Delli Gatti, D., Palestrini, A., Gaffeo, E., Giulioni, G., & Gallegati, M. (2008b). *Emergent macroeconomics: an agent-based approach to business fluctuations*. Heidelberg: Springer.

Delli Gatti, D., Gaffeo, E., & Gallegati, M. (2010a). Complex agent-based macroeconomics: A manifesto for a new paradigm. *Journal of Economic Interaction and Coordination, 5–2*, 111–135.

Delli Gatti, D., Gallegati, M., Greenwald, B., Russo, A., & Stiglitz, J. E. (2010b). The financial accelerator in an evolving credit network. *Journal of Economic Dynamics and Control, 34*(9), 1627–1650. https://ideas.repec.org/a/eee/dyncon/v34y2010i9p1627-1650.html.

Delli Gatti, D., Desiderio, S., Gaffeo, E., Cirillo, P., & Gallegati, M. (2011). *Macroeconomics from the bottom-up*. Heidelberg: Springer.

Delli Gatti, D., Gallegati, M., Greenwald, B. C., Russo, A., & Stiglitz, J. E. (2012). Mobility constraints, productivity trends, and extended crises. *Journal of Economic Behavior & Organization, 83*(3), 375–393. https://doi.org/10.1016/j.jebo.2012.05.01. https://ideas.repec.org/a/eee/jeborg/v83y2012i3p375-393.html.

Gallegati, M., & Palestrini, A. (2010). The complex behavior of firms size dynamics. *Journal of Economic Behavior & Organization, 75*(1), 69–76.

Gallegati, M., Palestrini, A., & Rosser, J. B. (2011). The period of financial distress in speculative markets: Interacting heterogeneous agents and financial constraints. *Macroeconomic Dynamics, 15*(1), 60–79.

Giri, F., Riccetti, L., Russo, A., & Gallegati, M. (2019). Monetary policy and large crises in a financial accelerator agent-based model. *Journal of Economic Behavior and Organization, 157*, 42–58.

Godley, W., & Lavoie, M. (2007). *Monetary economics an integrated approach to credit, money, income, production and wealth*. New York: Palgrave MacMillan.

Greenwald, B., & Stiglitz, J. E. (2003). *Towards a new paradigm for monetary economics*. Cambridge University Press. https://doi.org/10.1017/CBO9780511615207.

Greenwald, B. C., & Stiglitz, J. E. (1993). Financial market imperfections and business cycles. *The Quarterly Journal of Economics, 108*(1), 77–114.

Grilli, R., Tedeschi, G., & Gallegati, M. (2014). Bank interlinkages and macroeconomic stability. *International Review of Economics and Finance*. https://doi.org/10.1016/j.iref.2014.07.002.

Grilli, R., Tedeschi, G., & Gallegati, M. (2015a). Markets connectivity and financial contagion. *Journal of Economic Interaction and Coordination*. https://doi.org/10.1007/s11403-014-0129-1.

Grilli, R., Tedeschi, G., & Gallegati, M. (2015b). Network approach for detecting macroeconomic instability. In *Proceedings - 10th International Conference on Signal-Image Technology and Internet-Based Systems*, SITIS 2014. https://doi.org/10.1109/SITIS.2014.96.

Heemeijer, P., Hommes, C., Sonnemans, J., & Tuinstra, J. (2009). Price stability and volatility in markets with positive and negative expectations feedback: An experimental investigation. *Journal of Economic Dynamics and Control, 33*(5), 1052–1072.

Hommes, C., Sonnemans, J., Tuinstra, J., & Van de Velden, H. (2005). Coordination of expectations in asset pricing experiments. *Review of Financial Studies, 18*(3), 955–980.

Kindleberger, C. P. (2000). *Manias, panics, and crashes: a history of financial crisis* (4th ed.). New York: Wiley.

Leijonhufvud, A. (2006). Agent-based macro. *Handbook of Computational Economics, 2*, 1625–1637.

Marimon, R., Spear, S. E., & Sunder, S. (1993). Expectationally driven market volatility: An experimental study. *Journal of Economic Theory, 61*(1), 74–103.

Orcutt, G. H. (1957). *A new type of socio-economic system. The review of economics and statistics*, pp. 116–123

Palestrini, A. (2007). Analysis of industrial dynamics: A note on the relationship between firms' size and growth rate. *Economics Letters, 94*(3), 367–371.

Palestrini, A., & Gallegati, M. (2015). Unbiased adaptive expectation schemes. *Economics Bulletin, 35*(2), 1185–1190.

Recchioni, M. C., Tedeschi, G., & Gallegati, M. (2015). A calibration procedure for analyzing stock price dynamics in an agent-based framework. *Journal of Economic Dynamics and Control*. https://doi.org/10.1016/j.jedc.2015.08.003.

Riccetti, L., Russo, A., & Gallegati, M. (2013a). Leveraged network-based financial accelerator. *Journal of Economic Dynamics and Control, 37*(8), 1626–1640. https://doi.org/10.1016/j.jedc.2013.02.00. https://ideas.repec.org/a/eee/dyncon/v37y2013i8p1626-1640.html.

Riccetti, L., Russo, A., & Gallegati, M. (2013b). Unemployment benefits and financial leverage in an agent based macroeconomic model. *Economics - The Open-Access, Open-Assessment E-Journal, 7*, 1–44. https://ideas.repec.org/a/zbw/ifweej/201342.html.

Riccetti, L., Russo, A., & Gallegati, M. (2015). An agent based decentralized matching macroeconomic model. *Journal of Economic Interaction and Coordination, 10*(2), 305–322.

Riccetti, L., Russo, A., & Gallegati, M. (2016a). Financialisation and crisis in an agent based macroeconomic model. *Economic Modelling, 52*(PA), 162–172.

Riccetti, L., Russo, A., & Gallegati, M. (2016b). Stock market dynamics, leveraged network-based financial accelerator and monetary policy. *International Review of Economics & Finance, 43*(C), 509–524. https://doi.org/10.1016/j.iref.2015.10.03. https://ideas.repec.org/a/eee/reveco/v43y2016icp509-524.html.

Riccetti, L., Russo, A., & Gallegati, M. (2018). Financial regulation and endogenous macroeconomic crises. *Macroeconomic Dynamics, 22*(04), 896–930.

Russo, A., Catalano, M., Gaffeo, E., Gallegati, M., & Napoletano, M. (2007). Industrial dynamics, fiscal policy and R&D: Evidence from a computational experiment. *Journal of Economic Behavior and Organization, 64*(3–4), 426–447.

Russo, A., Riccetti, L., & Gallegati, M. (2016). Increasing inequality, consumer credit and financial fragility in an agent based macroeconomic model. *Journal of Evolutionary Economics, 26*(1), 25–47.

Schasfoort, J., Godin, A., Bezemer, D., Caiani, A., & Kinsella, S. (2018). Monetary policy transmission in a macroeconomic agent-based model. *Advances in Complex Systems, 25*, 1.

Schularick, M., & Taylor, A. M. (2012). Credit booms gone bust: Monetary policy, leverage cycles, and financial crises, 1870–2008. *American Economic Review, 102*(2), 1029–61.

Tedeschi, G., Iori, G., & Gallegati, M. (2009). The role of communication and imitation in limit order markets. *European Physical Journal B*. https://doi.org/10.1140/epjb/e2009-00337-6.

Tedeschi, G., Mazloumian, A., Gallegati, M., & Helbing, D. (2012). Bankruptcy cascades in interbank markets. *PLoS ONE*. https://doi.org/10.1371/journal.pone.0052749.

Yellen, J. (2011). Macroprudential supervision and monetary policy in the post-crisis world. *Business Economics, 46*(1), 3–12.

Economic History at the Faculty of Economics in Ancona

Francesco Chiapparino, Augusto Ciuffetti and Roberto Giulianelli

Abstract This contribution briefly describes the development of scientific research in economic history carried out at the Faculty of Economics "Giorgio Fuà", starting with Alberto Caracciolo, who is one of the founders of Italian post war historiography, and whose activity was carried on by other researchers who worked in Ancona, such as Sergio Anselmi, Ercole Sori, and Franco Amatori. The main topics of this research activity are described: they range from industrial history and the history of economic development, with their branches in business history, banking history, and the history of district economies, to the wide set of environmental and territorial historiographies—rural history, urban history, eco- and mountain history—as well as the history of the Adriatic Sea and of migration, etc. The purpose of this review is to show how the originality and innovation introduced by these studies have given Ancona University its prominent position in the Italian historiographic debate of the last fifty years.

1 Alberto Caracciolo and the Origins of the Faculty

In Ancona research into economic history has a short but rich and significant tradition which runs parallel to the development of the Faculty of Economics. In 1959, when the original nucleus of the Faculty was still part of the University of Urbino, Giorgio Fuà appointed Alberto Caracciolo to coordinate this scientific sector. At that time

F. Chiapparino (✉) · A. Ciuffetti · R. Giulianelli
Department of Economics and Social Sciences, Università Politecnica delle Marche, Piazzale Martelli 8, 60121 Ancona, Italy
e-mail: f.chiapparino@univpm.it

A. Ciuffetti
e-mail: a.ciuffetti@univpm.it

R. Giulianelli
e-mail: r.giulianelli@univpm.it

Caracciolo was a young and promising scholar who was to have a central role in Italian historiography in the following decades. Born in Livorno in 1926, Caracciolo studied under Federico Chabod and was an intellectual with a Gramscian–Marxist background. He was an active member of the Italian Communist Party from 1946 to the 1956 invasion of Hungary. Caracciolo remained in Ancona throughout the development of the new faculty, then moved to Macerata in 1970, to Perugia in 1972 and finally to Rome, where he taught at "La Sapienza" in the eighties and nineties. During this time, even thanks to his experience in Ancona, he had a crucial influence on the overall renewal of Italian historical studies, playing a decisive role at the same time in the education of a whole generation of researchers in this part of the country (Nenci 2005). His works and scientific initiatives allowed the "historical disciplines to emerge from the cocoon of old ethical-political historiography and open up to a new spectrum of knowledge, which also reflected the new cultural and civil wealth of Italian society" (Bevilacqua 2002). Together with Alberto Tenenti, Caracciolo was a protagonist in the introduction of the new methodological approaches which can be found in the *Annales* in Italy and the revolution that this produced in the historiographical culture of the country (Galasso 2002; Romanelli 2002). Important steps in the direction of opening up to new international trends were made by Caracciolo in Ancona (Sori 2011), starting from his dissertation on the 18th century *Port franc* (Caracciolo 1965), published in France, and leading up to the foundation of the review *Quaderni storici*, which not surprisingly was first published as *Quaderni storici delle Marche*, and the collective work on the development of industrial Italy (Caracciolo 1963), that—in line with Fuà's analyses—changed the historical perspectives of the origin of industrial development in the country. The following years saw Caracciolo's pioneering work on eighteenth century economy and on the history of the Papal State, also touching on areas such as industrial archeology, regional history, the history of the environment and of energy, just to mention some of the many research perspectives explored by such a tireless innovator (Sansa 2004).

All these topics, but even more so an attitude of open-mindedness towards new cultural tendencies and ideals, together with critical attention to the most innovative questions raised by the national and international historiographical debate, animated the works of those historians who worked more closely with Caracciolo in Ancona: Sergio Anselmi, Renzo Paci (who followed Caracciolo to Macerata), and Ercole Sori, as well as Franco Amatori, Marco Moroni, Patrizia Sabbatucci and other academics who studied in Ancona in those years. As evidence of this open-minded approach, which characterized the Faculty from its origins, the names of many other Italian and foreign scholars can be added to the list: Maurice Aymard, Peter Hertner, Maria Ines Barbero, Giorgio Porisini, Riccardo Faucci, Raul Merzario, Marzio Romani, Gianluigi Basini and Luciano Segreto, who also contributed at different periods of time to the development of economic history studies at the University in Ancona.

Starting from this common ground, these studies have branched out, during the last half century, into a rich economic and social historiography, ranging from industrial and business history to rural and environmental history, including the Adriatic Sea, migrations and local development, urban history and the history of welfare.

2 Industry and Enterprise

Business history and the history of industrial development are two of the most relevant research fields investigated by Faculty members. In the latter area, they have contributed significantly to the national debate with at least two works. In the first, Caracciolo, taking part in 1969 in the Fuà's pioneering work on Italian development, focuses on industrial mobilization during the First World War as fundamental for Italian economic growth. Thirty years later, Franco Amatori, while editing, together with Duccio Bigazzi, Renato Giannetti and Luciano Segreto, the volume of the Einaudi *Storia d'Italia* dedicated to manufacturing, produced a sound set of analyses on the identity and the characteristics of the national secondary sector (Amatori et al. 1999). Between and after these two contributions, Faculty members have been engaged in a wide range of works on the history of individual industrial sectors (Chiapparino 1997; Ciuffetti 2013; Giulianelli 2017) and the modernization of the Marche region in the 20th century, which is an interesting case of district development based on small and medium-sized industry (Sori 1987; Moroni 2008).

Even more significant is the impact of studies conducted in Ancona in the field of business history. The close collaboration with David Landes and Alfred D. Chandler jr. allowed Franco Amatori to introduce this peculiar approach to economic history in Italy, where before the eighties it was limited to rare (albeit excellent) works by scholars such as Franco Bonelli or Valerio Castronovo. This late start was the result of ideological prejudices (for a long time, in the post-war period, the labor movement and industry were preferred as a subject of study rather than enterprise and entrepreneurs) and methodological distrust (the firm seemed to be unsuitable for an approach based on macro-economic issues such as economic development, market dynamics and social processes). Following Chandler's approach and example (in addition to translating some of his main works), Amatori produced studies that are still considered milestones for Italian business history, such as his analysis of the types of national entrepreneurs (Amatori 1980), the history of the Rinascente (mass retailing; Amatori 1989) and that of Lancia cars (Amatori 1996). Later, this same area of research was also developed, often starting from different perspectives, by other historians in Ancona, as testified, among others, by the works of Sori (2005) on Merloni, Chiapparino and Covino (2007) on Perugina, Giulianelli (2012) on the Piaggio Group, and Ciuffetti (2017) on Aem.

3 The Adriatic, Migration and Urban Studies

One of the topics most frequently analyzed by historians at the Università Politecnica is the Adriatic. At a superficial glance this subject may appear to be only of local interest. However, inspired by the example of Fernand Braudel, *Les Annales* and Caracciolo, Sergio Anselmi showed how, between the Middle Ages and the modern age the Adriatic Sea was anything but a closed space, as its physical location would seem to suggest. A complex and multi-faceted area, the Adriatic acted as a bridge between East and West, being a reservoir for food resources, a crossroads for commercial transactions and, even more importantly, a mediator for cultural exchanges. Its strategic importance led to frequent conflicts for its control; it was the scene of opposing interests, often expressed with extreme cruelty, but it also enhanced the extraordinary affinities between the populations that have settled on its coasts throughout history. By underlining these inter-ethnic affinities Anselmi, and with him the wisest historiographers in the sector, began to speak of an Adriatic *koinè*, a common cultural sense of "belonging" which unites the two shores. Anselmi created a project regarding the Adriatic Sea that was not limited to the drafting of enlightening scientific contributions (Anselmi 1991a, b, 1994), but also led to the publication of three successful volumes of literary stories inspired by historical sources and real characters (Anselmi 1996, 1997, 2000a). Anselmi's project also promoted a lively academic environment which, once the *Quaderni storici delle Marche* had become *Quaderni storici* and had moved to Rome, led to the foundation of anew scientific journal called *Proposte e ricerche*. The latter—which was created in 1978 by Anselmi, Sori and Renzo Paci, and whose scientific and editorial board is still chaired by the Faculty of Economics in Ancona—has repeatedly returned to the subject of the Adriatic, both through published essays, and through a wider range of studies which have been hosted in its series of monographs. The Università Politecnica delle Marche has thus become an Italian and international point of reference for this field of investigation, which continues to be of interest, above all thanks to the research carried out by Moroni (2010, 2011, 2012; see also Canullo et al. 2011; Adriano and Cingolani 2018).

A further line of research that has characterized the work of historians in Ancona is linked to the origins and development in Italy of another important field of study: urban history. Ercole Sori was among the founders, in 1976, of the journal *Storia urbana* and one of the forerunners of an interdisciplinary approach for the analysis of urban spaces. It is within this framework that his studies on the mechanisms of social control implemented in Italian cities during the early modern age must be placed. In this perspective, the city becomes a sort of laboratory in which to analyze the different social and welfare control practices implemented by "total" institutions (hospitals, "lazarettos", mental asylums, hospitals, hospices, orphanages, prisons), as well as specific policies for public order, aimed at regulating systems of food rationing, labor, vagrancy and begging, etc. (Sori 1982). This innovative research activity developed in various directions. On one hand it evolved into participation in an EU Network of Excellence, on the subject of "Sustainable development in a diverse world", which allowed a more thorough historic perspective of topics linked to globalization and migration processes, such as social diversity and ethnic entrepreneurship (Chiapparino and Giulianelli 2010; Chiapparino 2011b, c). In these works, urban history issues, such as social diversity and marginality, are connected to another field of research which concerns Italian emigration between the nineteenth and twentieth centuries, in which Sori once again had a pioneering role with his 1979 study.

On the other hand, some researchers have addressed their attention to the model of the industrial city, the question of popular housing, the origin and subsequent transformation of factory paternalism which triggered the creation of workers villages and neighborhoods. In the latter field, the work of the historians at the Faculty in Ancona has also been extended to include industrial archeology, a field of study which encompasses urban history, environmental history and the history of production sites. The development of this line of research includes original studies (Chiapparino 2011a; Ciuffetti and Parisi 2012) and active participation in the management of the Associazione Italiana per il Patrimonio archeologico industriale and its journal *Industrial Heritage*.

Within the same macro-area, the specific sectors of social history, the history of assistance in the modern age, the welfare state and corporate welfare, have also stimulated relevant interest within the Faculty of Economics (Ciuffetti 2004, 2014; Ciuffetti et al. 2017). Thanks to the activation of degree courses for social workers, this research has also had an important role in the didactic choices of the Faculty.

4 Rural, Environment and Landscape History

Another central area of research developed in Ancona is the history of agriculture, with specific reference to the *Mezzadria* sharecropping system. This subject has been investigated at the UNIVPM through a long-term approach, going from the colonization of the late Middle Ages to its dissolution in the second half of the 20th century, in parallel with the development of widespread industrialization based on districts of small and medium-sized firms which are typical of the *modello marchigiano*. It is mainly thanks to Sergio Anselmi's thirty-year research engagement that it has been possible to understand the Mezzadria, not as a sort of vestige of the transition from feudalism to capitalism, but as a microcosm which is entirely functional to the economic and environmental equilibrium of rural society in the Marche and in other parts of Italy in the early modern age (Anselmi 2000b, 2001). More, Anselmi's work led to an extensive reconstruction of the world of Mezzadria in the Marche, which had never been explored with methodological precision and a comprehensive approach. Not surprisingly, this investigation, which went beyond agricultural history and involved the history of technology, material culture, anthropology and ethnology, led to the creation of the Museum of the Mezzadria in Senigallia, which was directed by Anselmi from its foundation in 1978 up to 2003.

Above all thanks to Marco Moroni, studies on agricultural history in the Marche have also developed in original directions, such as the analysis of farmers' multiple-activities and different forms of proto-industry (Moroni 2008; Ciuffetti and Parisi 2018), which represent an important aspect of the genesis of modern industrial districts, the history of agrarian education (Moroni 1999) and of the territory (Anselmi and Volpe 1987; Moroni 2003).

While Caracciolo was responsible for initiating research studies dedicated to environmental history with an essay that even today is a fundamental point of reference for researchers (Caracciolo 1988), it is Sergio Anselmi who continues along this path by giving the farming economy a strong ecological value. In his interpretation, sharecropping is not only functional to the characteristics of the rural landscape in the Marche but is also able to preserve it over time. Significantly, after the end of the *Mezzadria*, the agricultural land in the region has been affected by run-off phenomena and erosion processes, while the indiscriminate use of chemical fertilizers has contributed to the pollution of soils and aquifers. In terms of ecological history, a significant contribution also came from Ercole Sori with an unprecedented history of waste, considered as an integral part of production processes and constantly present in the economies of European cities from the Middle Ages to today (Sori 1999, 2001). This study made it possible to come to grips with an environmental issue that is constantly at the center of public debate on the management of urban spaces; it addresses not only the problem of urban hygiene, but also the management and disposal of waste, outlining the knowledge, techniques and administrative practices that have been implemented at each stage of the urbanization process.

Finally, once again thanks to Sergio Anselmi and later encouraged by Ercole Sori, a further and profitable area of ongoing research concerns the history of the societies and the economies of mountain areas, specifically in the Apennine area of Central Italy. This is a research topic of great interest which, starting from the conferences periodically organized by *Proposte e ricerche* during the eighties in Sestino (Arezzo), continues to the present day with recent analyses dedicated to Apennine economies, divided between forms of persistence and drives for change (Anselmi 1985; Antonietti 1989; Calafati and Sori 2004; Bettoni and Ciuffetti 2010). Currently, this branch of study is part of the research on internal areas promoted by the association RESpro (Network of historians for landscapes of production) and studies activated within the Department through a UNIVPM strategic project, both of which involve economic history Faculty members.

It is difficult to predict the directions that historical research will take in the future in Ancona: it is rather a paradox that historiography depends a lot on changes in the present, which can be extremely unpredictable. Nevertheless, it is reasonable to hope that, as in the past half century, the economic history researchers in the Faculty, despite their limited number, will be able to make a significant contribution to the historiographical and general cultural debate in Italy.

References

Adriano, P., & Cingolani, G. (2018). *Nationalism and terror. Ante Pavelic and Ustasha Terrorism from Fascism to Cold War*. Budapest: Ceu Press.

Amatori, F. (1980). Entrepreneurial typologies in the history of industrial Italy (1880–1960). *The Business History Review, 55*, 359–386.

Amatori, F. (1989). *Proprietà e direzione. La Rinascente, 1917–1969*. Milano: FrancoAngeli.

Amatori, F. (1996). *Impresa e mercato. Lancia 1906–1969*. Bologna: Il Mulino.

Amatori, F., Bigazzi, D., Giannetti, R., & Segreto, L. (Eds.). (1999). *Storia d'Italia. Annali 15. L'industria*. Torino: Einaudi.

Anselmi, S. (Ed.). (1985). *La Montagna tra Toscana e Marche Ambiente, territorio, cultura, economia, società dal medioevo al XIX secolo*. Milano: FrancoAngeli.

Anselmi, S. (1991a). *Adriatico. Studi di storia: secoli XIV-XIX*. Ancona: Clua.

Anselmi, S. (1991b). *Sette città jugo-slave tra Medioevo e Ottocento: Skopjie, Sarajevo, Belgrado, Zagabria, Cattigne, Lubiana, Zara*. Ancona: Quaderni di Proposte e ricerche.

Anselmi, S. (1996). *Storie di Adriatico*. Bologna: Il Mulino.

Anselmi, S. (1997). *Ultime storie di Adriatico*. Bologna: Il Mulino.

Anselmi, S. (2000a). *Mercanti, corsari, disperati e streghe*. Bologna: Il Mulino.

Anselmi, S. (2000b). *Chi ha letame non avrà mai fame. Studi di storia dell'agricoltura, 1975–1999*. Ancona: Quaderni di Proposte e ricerche.

Anselmi, S. (2001). *Agricoltura e mondo contadino*. Bologna: Il Mulino.

Anselmi, S., Volpe, G. (Eds.). (1987). *L'architettura popolare in Italia*. Roma-Bari: Laterza.

Anselmi, S., Di Vittorio, A., & Pierucci, P. (1994). *Ragusa (Dubrovnik) una Repubblica adriatica. Saggi di storia economica e finanziaria*. Bologna: Cisalpino.

Antonietti, A. (Ed.) (1989). *La montagna appenninica in età moderna. Risorse economiche e scambi commerciali*. Ancona: Quaderni di proposte e ricerche.

Bettoni, F., & Ciuffetti, A. (Eds.). (2010). *Energia e macchine. L'uso delle acque nell'Appennino centrale in età moderna e contemporanea*. Narni: Crace.

Bevilacqua, P. (2002). Alberto Caracciolo uno storico creativo. *La Repubblica*, November 21.

Calafati, A. G., & Sori, E. (Eds.). (2004). *Economie nel tempo. Persistenze e cambiamenti negli Appennini in età moderna*. Milano: FrancoAngeli.

Canullo, G., Chiapparino F. & Cingolani, G. (Eds.) (2011) *The Adriatic Balkan area from transition to integration*. Napoli: Esi.

Caracciolo, A. (Ed.). (1963). *La formazione dell'Italia industriale. Discussioni e ricerche*. Bari: Laterza.

Caracciolo, A. (1965). *Le port franc d'Ancone: croissance et impasse d'un milieu marchand au XVIIIe siècle*. Paris: S.E.V.P.E.N.

Caracciolo, A. (1969). La crescita e la trasformazione industriale durante la prima Guerra mondiale. In G. Fuà (Ed.), *Lo sviluppo economico in Italia. Storia dell'economia italiana negli ultimi cento anni*. Milano: FrancoAngeli.

Caracciolo, A. (1988). *L'ambiente come storia. Sondaggi e proposte di storiografia dell'ambiente*. Bologna: Il Mulino.

Chiapparino, F. (1997). *L'industria del cioccolato in Italia*. Bologna: Il Mulino.

Chiapparino, F. (Ed.). (2011a). *Il patrimonio industriale delle Marche*. Crace: Narni.

Chiapparino, F. (Ed.). (2011b). *The alien entrepreneur. Migrant entrepreneurship in Italian emigration (Late 19th–20th Cent.) and in the Immigration in Italy at the Turn of the 21st century*. Milano: FrancoAngeli.

Chiapparino, F. (Ed.). (2011c). *Diversità sociale e sostenibilità. Una prospettiva storica*. Bologna: Il Mulino.

Chiapparino, F., & Covino, R. (2007). *La fabbrica di Perugia. Perugina 1907–2007*. Perugia: Comune di Perugia.

Chiapparino, F., & Giulianelli, R. (2010). An historical perspective on sustainable diversity: Market and nation as catalysts of diversity in modern Europe (1800–1950). In Janssens, M. et al. (Eds.), *The sustainability of cultural diversity. Nations, cities and organizations* (pp. 32–54). Cheltenham: Edward Elgar.

Ciuffetti, A. (2004). *Difesa sociale. Difesa sociale. Povertà, assistenza e controllo in Italia, XVI-XX secolo*. Perugia: Morlacchi.

Ciuffetti, A. (2013). *Carta e stracci. Protoindustria e mercati nello Stato pontificio tra Sette e Ottocento*. Il Mulino: Bologna.

Ciuffetti, A. (2014). *L'assistenza come sistema. Dal controllo sociale agli apparati previdenziali: San Marino tra età moderna e contemporanea*. San Marino: Centro Sammarinese di studi storici.

Ciuffetti, A. (2017). *Il fattore umano dell'impresa L'Azienda elettrica municipale di Milano e il welfare aziendale nell'Italia del secondo dopoguerra*. Venezia: Marsilio.

Ciuffetti, A., & Parisi, R. (Eds.). (2012). *L'archeologia industriale in Italia Storie e storiografia (1978–2008)*. Milano: FrancoAngeli.

Ciuffetti, A., & Parisi, R. (Eds.). (2018). *Paesaggi italiani della protoindustria. Luoghi e processi della produzione dalla storia al recupero*. Roma: Carocci.

Ciuffetti, A., Trisoglio, F., & Varini, V. (Eds.). (2017). *Il welfare aziendale in Italia nel secondo dopoguerra. Riflessioni e testimonianze*. Milano: Egea.

Galasso, G. (2002). Alberto Caracciolo. Da Marx alle Annales. *Corriere della Sera*. 21 November.

Giulianelli, R. (2012). *I Piaggio La parabola di un grande Gruppo armatoriale e cantieristico italiano (1875–1972)*. Bologna: Il Mulino.

Giulianelli, R. (2017). *Armatori, banche e Stato. Il credito navale in Italia dall'Unità alla prima crisi petrolifera*. Bologna: Il Mulino.

Moroni, M. (1999). *Istruzione agraria e sviluppo agricolo nelle Marche dell'Ottocento*. Ancona: Quaderni di Proposte e ricerche.

Moroni, M. (2003). *L'Italia delle colline. Uomini, terre e paesaggi nell'Italia centrale (secoli XV-XX)*. Ancona: Quaderni di Proposte e ricerche.

Moroni, M. (2008). *Alle origini dello sviluppo locale. Le radici storiche della Terza Italia*. Bologna: Il Mulino.

Moroni, M. (2010). *Tra le due sponde dell'Adriatico. Rapporti economici, culturali e devozionali in età moderna*. Napoli: Esi.

Moroni, M. (2011). *L'impero di San Biagio. Ragusa e i commerci balcanici dopo la conquista turca (1521–1620)*. Bologna: Il Mulino.

Moroni, M. (2012). *Nel medio Adriatico. Risorse, traffici, città fra basso medioevo ed età moderna*. Napoli: Esi.

Nenci, G. (Ed.). (2005). *Alberto Caracciolo uno storico europeo*. Bologna: Il Mulino.

Romanelli, R. (2002). E anche l'Italia ebbe le sue "Annales". *Il Sole 24 Ore*. 24 November.

Sansa, R. (2004). Bibliografia degli scritti di Alberto Caracciolo. *Quaderni storici, 115*, 279–303.

Sori, E. (1979). *L'emigrazione italiana dall'Unità alla seconda guerra mondiale*. Bologna: Il Mulino.

Sori, E. (1982). *Città e controllo sociale in Italia tra XVIII e XIX secolo*. Milano: FrancoAngeli.

Sori, E. (1987). Dalla manifattura all'industria (1861–1940). In S. Anselmi (Ed.), *Le Marche*. Torino: Einaudi.

Sor, E. (1999). *Il rovescio della produzione. I rifiuti in età pre-industriale e paleotecnica*. Bologna: Il Mulino.

Sori, E. (2001). *La città e i rifiuti. Ecologia urbana dal Medioevo al primo Novecento*. Bologna: Il Mulino.

Sori, E. (2005). *Merloni. Da Fabriano al mondo*. Milano: Egea.

Sori, E. (2011). Alberto Caracciolo ad Ancona: l'insegnamento e le ricerche di storia economica nella Facoltà anconitana e nelle Marche. In E. Sori & R. Giulianelli (Eds.), *Consumi e dinamiche economiche in età moderna e contemporanea*. Napoli: Esi.

Sociology in the Faculty of Economics

Ugo Ascoli, Carlo Carboni and Giovanna Vicarelli

Abstract The initial presence of Alessandro Pizzorno and Massimo Paci, the founding fathers of General Sociology and Economic Sociology in Italy, has given young students and those graduating from the Economics Faculty the opportunity to embark enthusiastically on research in the fields of sociology. In the beginning, studies centred on the economic development of small and medium size firms and on its interaction with family and parental networks; attention then shifted towards the social and environmental cost of these forms of economic growth and hence to public policies. This triggered the interest in the welfare system, in its actors and the various social policies such as those regarding work, pensions, health, social welfare services, education and the home. Economic development and the welfare state also have a profound influence on inequality structures and social classes both of which have become the focus of important in-depth analyses. Meanwhile the study of health policies and healthcare professionalism was boosted by the setting up of the interdepartmental Research Centre on Integration in Social and Health Services (CRISS). Other lines of research have investigated the ruling classes and the new form of civil and social mobilization of the current period.

U. Ascoli · C. Carboni · G. Vicarelli (✉)
Department of Economics and Social Sciences, Università Politecnica delle Marche, Piazzale Martelli, 8, Ancona, Italy
e-mail: m.g.vicarelli@staff.univpm.it

U. Ascoli
e-mail: u.ascoli@univpm.it

C. Carboni
e-mail: c.carboni@univpm.it

S. Longhi et al. (eds.), *The First Outstanding 50 Years of "Università Politecnica delle Marche"*,
https://doi.org/10.1007/978-3-030-33879-4_4

1   The Sixties and Seventies: Research on Industrialization and the Transformation of the Italian Labour Market

At the end of the Sixties, Italy underwent a period of social mobilization that is unmatched in the history of the Republic. It lasted a decade, thus spanning a time frame which is much longer than the cycles of protest seen in the same years in France, Germany and other European countries. The Sociology School in Ancona grew from research on student and worker struggles; from there it was a short step to the examination of what was changing in the labour market.

Reflection and research were focused on the prevalent forms of industrialization adopted in the Marche, based on small and medium size enterprises, on an impressive network of artisan firms and on the ample availability of a workforce both in the factory and at home (Bugarini and Vicarelli 1978). In this way, the image of a 'Third Italy' took shape in which the Marche was grouped together with the other regions of Central Italy and the North East. It was an industrial development of enormous significance based on small units of production and the division of labour among firms in the so-called 'industrial districts' (Ascoli 1975). This was a part of Italy that was growing at a very fast rate when compared with the North West where the growth of big firms was slowing down, and with the South where no industrial fabric worthy of that name had been developed around state industries, commonly called 'cathedrals in the desert'.

Under the excellent guidance of Massimo Paci, the young sociologists in Ancona identified numerous relationships between family kinship networks and forms of work ranging from 'regular', more or less trade unionized work in a factory, to work at home and also in the black market (Paci 1980). Low labour costs, high tax evasion levels, total workforce flexibility with the involvement of the entire family unit that followed precise 'gender rules' (work in the firm was for men and young women; work at home was for young mothers and adult women) coupled with an extraordinary entrepreneurial ability enabled the Marche, in just a few years, to make the happy and fortunate shift from a share-cropping region to one that was highly industrialized. At the same time, the first potential bottlenecks and 'social' cost of these types of growth were becoming evident, namely the effects on workers' health and on the environment, the need for social services to take care of those who were not able to keep pace and on women's growing awareness that they wanted to free themselves from an extremely paternalistic and male chauvinist regime.

Meanwhile social mobilization which included workers' struggles, student protests, the fight for civil rights, the explosion of feminism and the protest by the Studenti Medi (high-school students) had a profound effect on party politics and trade unions. On the one hand, the Seventies brought about the most important manifestation of an awakening of civil society in the areas of protection and social well-being and on the other, fundamental policy innovations. It was not without reason that the decade closed with the establishment of the National Health Service (1978). Even personal services had seen important legislative changes, for example, to choose one of many, the law that set up municipal playschools (1971). Pensions, for their part, had undergone significant reform at the end of the Sixties (1969) which greatly improved the pension prospects for salaried workers and a network had been created to alleviate poverty for the elderly. Politics showed the will to take on the most important social problems in the country, albeit in different ways: the universalistic approach to health and schools and the particularistic approach in the pension and work sectors (Ascoli 1984).

2   The Eighties and Nineties: The First Analyses of Italian Welfare in the European Framework and of Changes in the Social Structure

In the Eighties and Nineties the areas of academic interest slowly shifted from the analysis of structural transformations in the economy and Italian society to the study of institutions and public policies.

Some areas of research remained the same, namely male and female migration, the transformation of the family in the Marche and the roles of the members therein (David and Vicarelli 1983, 1994; Moretti and Vicarelli 1986; Vicarelli 1994). Furthermore, research on the job market and local development led inevitably to interest in the interpretation of the Italian social structure and its change over time (Paci 1982, 1991, 1992; Ascoli and Catanzaro 1987). The focal point of the research was both the growth of the working class, one of the chief players in industrialization (Fordist and widespread), and also the entrepreneurial manufacturing middle class that appeared to guide the development of the Third Italy. Alessandro Pizzorno was one of the fathers of the Ancona sociologists and it was his valuable work that bridged the gap not only between the analysis of social composition and the redistributive and fiscal public policies but also between the socio-economic and socio-political dimensions (Pizzorno 1974). This shone the spotlight on the middle classes which, although multifaceted and heterogeneous, were at the same time, the fulcrum of the new industrial society and keystone of political consensus (Carboni 1981). It was also Massimo Paci's belief that the middle classes played a central role in the Italian social structure, demonstrated both by their very strong growth in urban areas and as salaried workers and by the tenacity and resilience of the independent middle classes which, along with artisans and small business owners, were the leading players in the widespread industrialization (Novelli 1990; Paci 1992).

However, the majority of research was oriented towards welfare policies, focusing on the actors who presided over them, the relation between public and private and the comparison between the main European countries. The research showed the specific form assumed by the Italian system within the European welfare model: little space allotted to services and more emphasis on money transfers. On balance then, a particularistic approach and one of patronage in the planning and distribution of services, a weighty devolution of responsibility to families as well as the strong and growing importance of organized voluntary work, co-operatives and associations in general

(Ascoli 1984). The market appeared to provide welfare tools that were effective only for the elite and upper middle classes. Policies in the southern regions were revealed as being less able to cope with the principal social risks; the social dualization process was already materializing with increasing momentum, a circumstance which in the following years was to lead to a situation at odds with the idea of national welfare, referring instead to 'northern welfare' and 'southern welfare' (Ascoli 2011).

The Welfare State, in fact, emerged as a complex structure to which factors of varying nature contributed. These conformed to both the ideological and moral tendencies (altruism, egotism, etc.) found in society and to the principal forms of social aggregation (in particular classes and organized groups) as well as what were really more political factors (role of the parties, governments, etc.). Moreover, an awareness was gaining ground that it was necessary to study and analyse 'long-term processes' in order to understand the continual oscillation between the different possible combinations of types of social protection (state, market, community) (Paci 1982, 1989). Alongside the State and political and administrative elite other key players appeared. These were not just the parties and the big trade unions but also professional bodies, voluntary organizations, insurance companies, social movements, informal mutual aid groups, big class groups and the new emerging classes, the family and the family network (Ascoli 1985, 1987; Ascoli and Pasquinelli 1993; Ascoli and Pavolini 1999; Ascoli and Ranci 2002, 2003).

In those years, the welfare policies that had developed in Europe from the latter half of the nineteenth century became the object of a growing body of studies and research which led both to the accumulation of a great deal of quantitative and qualitative comparative data and to the elaboration of numerous explanatory theories. In general terms, there were two approaches that typified the explanations of the welfare state. On the one hand, there were the living conditions resulting from industrialization and urbanization and this was the case whether pluralistic or Marxist theories were used to explain the new social policies. On the other, the necessity for social needs to be interpreted and legitimized so as to become the object of claims and focused policies. With regard to this debate, the research by the Ancona sociologists was carried out with a view to integrating the two approaches and principally the interconnection between different mechanisms for regulating politics, the economy and social issues (Ascoli 1986, 1988, 1992a, b). This meant recognizing that cultural factors and social movements (including professional movements) played an appreciable role in the definition of the emerging models of welfare.

It was this aspect that became the subject for the long-term analysis of Italian healthcare policies (Vicarelli 1986, 1997a, b). The intention was to highlight, in the process of the creation and development of the National Health Service, the role played by some at risk categories (women, children, the elderly, mentally ill etc.) and the intermediate social classes (doctors, social workers, self-employed lower middle class) who had at times found the healthcare policy advantageous and at times disadvantageous and for this reason chose to support it or forgo it. With this perspective in mind, the aim was to identify the thread which linked late nineteenth century healthcare policies with those of the twentieth century in order to be able to reflect on the present. In fact, the delay in Italy in starting to pinpoint policies in support of the physical and material well-being of the population as well the forms that were employed seemed to play a determining role in the configuration of the successive welfare system. The space set aside for private charities, the role constantly attributed to family solidarity, the professional power long denied to doctors, the political dominance of some of the big public and semi-public organizations, the growing territorial and social inequalities in the distribution and access to services and the participatory but also fragmented zeal of the civil society were some of the features that are evident right from the initial phases of development of the healthcare policies but that are repeated in the following phases typified by the slowness in drafting laws and the difficulty of enforcing them (Vicarelli 1988, 1989; David 1989). On the whole, these were theoretical hypotheses that grew out of a long work of documentation and empirical research by some sociologists of the Ancona school and which proved the importance of cultural and social factors in explaining the welfare state. Consequently, in academic debate it was deemed necessary to express a sociocultural viewpoint to support and integrate with those of the more widespread economic-structural and political-institutional matrix.

This research also triggered interest in the role played by some social classes (among which, that of doctors) whose professional growth was tied to the transformative dynamics of the welfare systems. The international debate on the processes of professional development were characterized, therefore, by the presence of two widely recognized types of professionalization: the Anglo-Saxon and the continental. In the former, professional groups that initially worked on the market had succeeded in obtaining the monopoly of their profession from the State thus keeping extensive autonomy and control over their working conditions. In the latter, it was the bureaucratic hierarchies that had transformed themselves into professions driven by the acquisition of academic qualifications and that had challenged the power and elevated social status of the dominant aristocratic groups. Whereas the research carried out by the Ancona school tended to show that in Italy a third professionalization model seemed to be developing centred on clan identity rather than on the predominant role of the market or state. At least for the medical profession, in a polycentric system like the insurance and corporate schemes that had been created in the Fascist years and been re-proposed after the war, the opportunity of having a position of control could only come about by being able to be present simultaneously in the various health facilities (hospital, specialized healthcare and that of the GPs) supplying each with limited economic resources, prestige or power. However, this sum total of professional positions could prove to be dispersive or ambiguous in terms of social identity and political representation if not supported by solid grids of membership originating from family and parental networks, scientific-academic networks and the more or less legal associational networks of masonry and territorial mafia. Following this line of reasoning, it was therefore possible to affirm that the professions represented a fourth and distinct institutional basis of the social order alongside the State, the Market and the Community. It was thus able to provide a long-term autonomous contribution to making the performance of the social players reciprocally flexible and predictable. In Italy, examples supporting this approach were to be found in specific sectors of the medical profession such as in the surgeries of GPs. Moreover, the teachings of Elias were seen as theoretical and empirical bases worth re-evaluating in a scientific context like that of Italy which was poorly disposed to both the study of historical sociology and relational interdependences.

3   The Twenty-First Century: From the Study of Individual Welfare Policies to New Thinking on the Structural Features of Italian Society

At the end of the Nineties, the chief achievements of the Ancona sociologists on the characteristics of the welfare system became a reference point for the debate in the country on the future of social protection policies. Since then, deeper analysis has been focused, on the one hand, on individual welfare policies in healthcare, personal social services, pensions, work, education and housing sectors and on the other, the role of voluntary work and associations in the planning and provision of welfare measures Pattarin 2005, 2011; Moretti 2008; Ascoli 2011; Ascoli and Pavolini 2012; Giarelli et al. 2012). At the same time, one line of research developed which was directed at analyzing the characteristics of the elite and ruling classes, underlining the peculiarities with regard to other European countries such as the role they played in the national political system.

It is within this framework that some sociologists of the Economics Faculty have focused their research on the Sociology of Health and Medicine (Vicarelli 2013; Ingrosso and Vicarelli 2015), the foundation of which as a specific section of the Italian Association of Sociology took place here in the Ancona branch (2005), one of whose members was the first national president. At the same time, the creation of an interdepartmental research centre to study the themes of healthcare and welfare (CRISS) has created the institutional context for carrying out a variety of studies on emerging subject areas (Vicarelli 2005, 2011). In the following two decades these have become the basis for a wide range of specialized courses and university Master degrees for employees of the National Health System on the subjects of health management in the new organizational networks. Furthermore, in the academic year 1990/1, at the Economics Faculty, university courses in the social services sector were launched in which the sociologists have played an extensive role in the planning and institutional responsibility as well as in the lecturing and research. The curriculum offered includes the Specialization School, the three-year undergraduate degree in Social Services and the Master in Organization and Management of the Social Services.

Since the early 2000s, a line of research has been developing which concerns the professionalization of the non-medical healthcare occupations (Pattarin 2009). In healthcare systems that were undergoing a phase of profound change, processes of requalification and socio professional relocation of various healthcare jobs such as that of obstetricians, nurses and social workers became part of the debate on the crisis in medical dominance (Vicarelli 2001, 2007; Spina 2009; Moretti et al. 2012;

Vicarelli and Spina 2015). The study on obstetricians, for example, carried out by the Ancona School from a historical-comparative viewpoint, aimed to understand if and to what extent the process of professionalization, which had taken place on a formal level through a long regulatory process, had repercussions on the practical level thus determining the professional growth of this category. The new professional position of social workers has also become the object of study with the aim of demonstrating the specific features (for example, in relation to more vulnerable social occupations like educators or nursing aides) and the possible new location in the network of social services for example, in GPs' surgeries. At the same time, the reconstruction of the processes of the creation and development of the medical profession for women (Vicarelli 2003, 2008; Spina and Vicarelli 2015; Vicarelli and Bronzini 2018a) or what characterized the young doctors from the Marche tended to demonstrate the way in which the welfare policies were closely interwoven with generational changes and gender (Spina 2017). This led to structural and subjective conditions that differed over time for the very profession which until then had been dominant (Vicarelli 2000, 2006, 2010a, b; Bronzini 2004; Speranza et al. 2008).

It was the combination of these studies that led members of the Ancona group to take a new interest in the sociology of professions in Italy as they recognized the need to 'rethink the foundations. They not only observed the European debate on hybrid forms and new professionalism, but were also inspired by the theoretical thought of Elias. This resulted in several lines of research: on the possible hybridization of the medical profession following the privatization of the Health Service (doctor–managers) (Vicarelli and Pavolini 2017), on the transformation of dentists in the new contexts of economic crisis (Vicarelli and Spina 2015), and on the influence of the New Public Management with regard to the teaching profession (Bronzini and Spina 2018).

In this period, one of the emerging topics of research was the question of social inequalities in health (Bronzini 2009). Several international studies, starting from the beginning of the Nineties, attested that the extension of the forms of health protection, the progress made in medicine and increased life expectancy went hand in hand with a growing inequality in health among social groups. The work of the Ancona sociologists enter this debate with a theoretical contribution aimed at demonstrating the way in which the social distance between individuals based on gender, socio-economic class, working conditions, cultural group and not least, the territorial context all reflect on actual and perceived health conditions but also on the actual demand for health, on the possibility of access to the provision of healthcare services and on the consumption of those services (Moretti 2016, 2017a).

Moreover, the changes in the epidemiological and demographic scenario, the spread of chronic illnesses and the transformation in family structures in the new contexts of the economic crisis of the early years of this century seemed to call for a re-composition of the traditional split between formal and informal curative pathways as well as between professional knowledge and that of the layperson (Vicarelli and Bronzini 2009; Bronzini 2016; Vicarelli 2016). These scenarios opened the way to new reflections on the need for the co-involvement of citizens, both in the fostering of the quality of life in relatively illness-free periods and in the therapeutic pathways especially those dealing with chronic cases. Hence, on the one hand, the necessity to rebuild and re-contextualize the sociological debate on the role of the patient within the social sciences and on the other, to embark on the study of some practical lines of action (Moretti 2011). These ranged from the institutionally recognized voice, the citizens in the regulatory context of the Italian National Health Service, the active commitment of families in the processes of horizontal subsidiarity and lastly to covering specific cases of activism expressed by patients and organizations often with the support of the more involved social and healthcare professions.

This line of study was reflected in the creation of a model of "Therapeutic education of patients and their family care-givers" (Family Learning Socio-Sanitario—FLSS) which has been adopted for various chronic diseases. On this issue, the Ancona school created a specific pathway 'the Third Mission' which, in close collaboration with the territorial and hospital services of the Marche Regional authority, has contributed to developing self-care skills of families and patients as well as increasing their empowerment. Most recently, starting from these lines of research and training, some of the sociologists of the Economics Faculty have turned their attention to the influence of Information and Communications Technology (ICT) both regarding citizens' demands for healthcare and for professional and organizational responses given within the National Health Service and the new private care markets (Vicarelli and Bronzini 2009, 2018b).

As a whole, the research in the different thematic areas has led to establishing that in the last twenty years there has been a tendency to reduce the range of public actions of the so-called Social Welfare (Ascoli 2011; Ascoli and Pavolini 2015): faced with new social risks, Italian Welfare has appeared incapable of recalibration in order to effectively tackle the new challenges which range from chronic illness to disability, from the need to find a balance between private and working life, to the necessity of reorganizing the transition from school to the job market (Vicarelli 2015). To this can be added the many aspects of the new poverty, long-term unemployment, migratory movement from the 'south' of the world and new housing emergencies. The prospect, for example, of making Italians a nation of house owners, nurtured since the great post war reconstruction plans, has contributed to the weak development of housing welfare. Yet from the beginning of this century, first, the speculative house market bubble and secondly, the prolonged economic crisis have underlined that the 'housing question' was anything but resolved for a swathe of population (on the increase) whose access to the housing market remains barred. This is the background then to the line of research on housing policies carried out by the Ancona school. The work was organized on two fronts: on the one hand, the in-depth study of public housing and the problematic issues associated with it, on the other, the study of the most recent experiences of social hosing (Bronzini 2014, 2017; Moretti 2015, 2017b, c). Moreover, the last few years have seen the start of significant privatization processes, families have been burdened with increasingly heavy charges and the divide between southern and northern welfare has widened. Within the individual sectors, the area covered by voluntary work and organizations has diminished while the role of non-profit companies and cooperatives, but above all, of for-profit companies and the market in general has increased.

Some Ancona sociologists then turned their attention towards focusing on the forms of welfare connected with employment that were escalating in importance, the so-called occupational welfare, namely pension funds and health benefits; the provision in firms of personal social services ranging from nursery schools to other welfare measures for reconciling work with personal and family issues (Pavolini et al. 2013).

The Ancona sociologists were the first in Italy to understand the importance of this aspect of welfare and through the first national survey of the topic in 2012, they succeeded in measuring its spread and growth trend, while at the same time, emphasizing the advantages and disadvantages regarding universalist cultures and citizenship, as well as the processes of social differentiation. They underlined the risk of a return to forms of social protection based not, as formerly, on social citizenship but on the type of job, where there is an increasing divide between the employed and unemployed, between workers in medium size and big companies and workers in small and tiny companies, between employees with permanent contracts (typical) and employees with fixed term contracts (atypical), between the communities of the Centre North and South, and lastly between private and public sector employees. In the South and in the public sector occupational welfare barely exists. In the face of the Great Recession that engulfed the country for almost a decade, a clear trade-off has taken shape between wage restraint policies and the development of the so-called enterprise-based welfare: pension funds however, struggle to take off while health funds have seen strong growth to the point where they involve almost half of private employees (Agostini and Ascoli 2014; Ascoli and Pavolini 2015; Ascoli et al. 2018). The critical state of the National Health Service has meant that workers look very favourably on substitute healthcare financed mainly by firms (Vicarelli 2015). In the light of the healthcare actually provided by health funds and of the eventuality of 'substitution', research in general has fuelled great perplexity regarding the possible reshaping of the universalist service, while the sociology department in Ancona has clarified possible ways of reorganizing and strengthening the public service (Arlotti et al. 2017).

Occupational welfare has grown in importance fostered largely by the State through generous tax breaks, but also because of the increasingly glaring shortcomings of public Social Welfare. Following the intuition of Richard Titmuss, some of the Ancona sociologists turned their attention to researching the so-called fiscal welfare, that is, the missing state revenues, tax expenditures, which were the consequence of precise political decisions aimed at favouring certain categories or of supporting certain policy decisions (Pavolini and Ascoli 2019). Tax expenditures linked to welfare alone, according to some estimates, amounted in 2107, to forty-seven billion euros. Nor should it be forgotten that tax evasion has for years annually subtracted over one hundred billion euros from the State coffers. Recent and very recent history shows how the Italian welfare system is increasingly refining its 'model' based on money transfers and tax breaks. The resources earmarked for the big universalist service systems, ranging from health to education, continue not to grow, in fact they are diminishing. However a succession of measures are directed at improving the conditions of certain categories of pensioners, and a timid attempt to tackle poverty is taking shape. At the same time, through tax exemption mechanisms laid down in the recent stability laws, the so-called 'enterprise-based welfare' has been further strengthened; firms are pushed to transforming themselves into 'mini welfare states', providing their employees with supplementary pension and health protection on top of what they are entitled to from the state system. They also provide personal services which are scarcely or not available at all in the territories.

This process of transformation and evolution of the welfare system poses a series of important questions on the effects this can have on the structure of the inequalities in a country that is seeing a steady growth of the numbers of people in poverty (Ascoli et al. 2015; Ascoli and Pavolini 2017). Social inequalities have been growing furiously in the last few years: increasingly significant splits in labour market participation are appearing not only between the territories but also between gender, age and nationality (Orazi 2017). The country seems to be divided into two between Centre-north and South: access to the universalist healthcare and education is becoming problematic for the weaker classes.

While reflecting on the large social protection systems and on the role of the leading players both public and private, attention could not but also be drawn to the role of the ruling classes. And this is what some of the Ancona sociologists did. With work behind them on the middle classes and classes in general, classes and citizenship, they started from the simple observation of citizens' growing degree of distrust of the elite, particularly those actually elected, the choice of which seems less and less to depend on merit and increasingly on devotion to the leaders (Carboni 2000, 2007a, b, 2015a, b). A theory tested also at the comparative level (Carboni 2015c). The present disenchantment and social cynicism as well as the loss of political authority lie in the fact that citizens cannot rely on the preferences of the politicians who are concerned not with public policies which meet the preferences of citizens, but their own individual interests, above all the interest in being re-elected (Carboni 2008). From this point of view, Italy seems like a "failed country" also because the ruling classes have not known how to guide it but have rather followed their own self-interest. Fifty years of studies by the Ancona sociologists on Italian society, its structure (Carboni 2002), change and ruling classes have also kept interest alive on another borderline issue, such consumerism, solidarity-based economics (Orazi 2013, 2018) and social movements. In recent years, some sociologists in the group have studied 'grillismo', beginning with the setting up of the 'meetup' to the current Five Star Movement (Orazi and Socci 2014). Others have examined the so-called Italian 'generation gap' and the duty of public policies to cope with it (Carboni et al. 2017).

All the results of the Ancona group's research and those of sociology nationally demonstrate that a change of direction is necessary, aimed principally at: bolstering education, training and research; increasing the efficiency and effectiveness of healthcare provision, trying to close the growing divide between the southern regions and the rest of the country; recovering public resources by tackling tax evasion, so as to provide the country with robust personal social services ranging from childcare to care of the disabled and the integration of migrants. These actions, if they were promoted by a ruling class attentive to the 'public good', would contribute substantially to creating new employment, especially for the young and for women. Management of large systems of social protection should remain in public hands in a project that is able to best utilize the third sector resources (particularly voluntary work) and the market in order to enhance and expand the system and meet the needs of families with committed policies aimed at supporting parenting and care work. To this end, the Ancona group has recently developed a stream of research on how public administrations and services deal with pressures 'from above' aimed at adopting a managerial logic, 'from below' claiming more tailored services (Coletto and Bronzini 2018), and from the context to meet changing social problems.

References

Agostini, C., & Ascoli, U. (2014). Il Welfare occupazionale: un'occasione per la ricalibratura del modello italiano? *Politiche Sociali, 2,* 263–280. https://doi.org/10.7389/77343.

Arlotti, M., Ascoli, U., & Pavolini, E. (2017). Fondi sanitari e policy drift. Una trasformazione strutturale nel Sistema Sanitario Nazionale italiano? *RPS, 2,* 77–92.

Ascoli, U. (1975). Dispersione produttiva e occupazione precaria nelle Marche. *Inchiesta 5*(20).

Ascoli, U. (Ed.). (1984). *Welfare state all'italiana.* Bari: Laterza.

Ascoli, U. (1985). Welfare State e azione volontaria. *Stato e Mercato, 13*(1), 111–158.

Ascoli, U. (1986). The Italian welfare state between incrementalism and rationalization. In L. Balbo & H. Nowotny (Eds.), *Time to care in tomorrow's welfare systems: The Nordic experience and the Italian case* (pp. 107–141). Vienna: European Centre for Social Welfare Training and Research.

Ascoli, U. (Ed.). (1987). *Azione volontaria e Welfare State.* Bologna: Il Mulino.

Ascoli, U. (1988). The Italian welfare system in the 80's: Less state and more market? In R. Morris (Ed.), *Testing the limits: International perspectives on social welfare change in nine countries* (pp. 165–192). Hanover, NH: University Press of New England.

Ascoli, U. (1992a). L'azione volontaria nei sistemi di Welfare. *Polis, 6*(3), 429–436.

Ascoli, U. (1992b). Towards a partnership between statutory sector and voluntary action? Italian welfare pluralism in the '90's". In S. Kuhnle & P. Selle (Eds.), *Government and voluntary organizations* (pp. 136–156). Aldershot: Avebury.

Ascoli, U. (Ed.). (2011). *Il welfare in Italia.* Bologna: Il Mulino.

Ascoli, U., & Catanzaro, R. (Eds.). (1987). *La società italiana degli anni '80.* Bari: Laterza.

Ascoli, U., Natali, D., & Pavolini, E. (2018). Still a weak occupational welfare in Southern Europe? Evidence from the Italian case. *Social Policy Administration, 52,* 534–548. https://doi.org/10.1111/spol.12382.

Ascoli, U., & Pasquinelli, S. (Eds.). (1993). *Il Welfare mix. Stato sociale e terzo settore.* Milano: Angeli.

Ascoli, U., & Pavolini, E. (1999). Le organizzazioni di terzo settore nelle politiche socio-assistenziali in Europa: realtà diverse a confronto. *Stato e Mercato, 57*(3), 441–475.

Ascoli, U., & Pavolini, E. (2012). Ombre rosse. Il sistema di welfare italiano dopo venti anni di riforme. *Stato e Mercato, 96*(3), 429–464. https://doi.org/10.1425/38645.

Ascoli, U., & Pavolini, E. (Eds.). (2015). *The Italian welfare state in a European perspective.* Bristol: Policy Press.

Ascoli, U., & Pavolini, E. (Eds.). (2017). *Volontariato e innovazione sociale oggi in Italia.* Bologna: Il Mulino.

Ascoli, U., & Ranci, C. (Eds.). (2002). *Dilemmas of the Welfare mix.* New York: Kluwer.

Ascoli, U., & Ranci, C. (Eds.). (2003). *Il Welfare Mix in Europa.* Roma: Carocci.

Ascoli, U., Ranci, C., & Sgritta, G. B. (2015). *Investire nel sociale: La difficile innovazione del welfare italiano.* Bologna: Il Mulino.

Bronzini, M. (2004). Quanti sono i medici in Italia? *Una difficile risposta. Salute e Società, 3*(1), 145–160.

Bronzini, M. (Ed.). (2009). *Sistemi sanitari e politiche contro le disuguaglianze di salute.* Milano: Angeli.

Bronzini, M. (2014). *Nuove forme dell'abitare: l'housing sociale in Italia.* Roma: Carocci.

Bronzini, M. (Ed.). (2016). *Vissuti di malattia e percorsi di cura. La sclerosi multipla raccontata dai protagonisti.* Bologna: Il Mulino.

Bronzini, M. (2017). Contested issues surrounding social sustainability and self-building in Italy. *International Journal of Housing Policy, 17*(3), 353–373. https://doi.org/10.1080/14616718. 2016.1223450.

Bronzini, M., & Spina, E. (2018). Italian teachers: A profession in transition? *Cambio, 8*(16), 83–98. https://doi.org/10.13128/cambio-23255.

Bugarini, F., & Vicarelli, G. (1978). Interazione e sostegno parentale in ambiente urbano. *Rassegna Italiana di Sociologia, 3,* 464–493.

Carboni, C. (Ed.). (1981). *I ceti medi in Italia.* Roma-Bari: Laterza.

Carboni, C. (Ed.). (2000). *Le power élites in Italia. Chi conta nella società della comunicazione.* Roma: Ediesse.

Carboni, C. (2002). *La nuova società. Il caso italiano.* Roma-Bari: Laterza.

Carboni, C. (2007a). Verso un'analisi della ristratificazione sociale. *RPS, 4,* 77–90.

Carboni, C. (Ed.). (2007b). *Élite e classi dirigenti in Italia.* Roma-Bari: Laterza.

Carboni, C. (2008). *La società cinica. Le classi dirigenti italiane nel tempo dell'antipolitica.* Roma-Bari: Laterza.

Carboni, C. (2015a). Il ceto politico locale e regionale. In M. Salvati & L. Sciolla (Eds.), *L'Italia e le sue regioni* (Vol. I, pp. 251–270). Roma: Treccani.

Carboni, C. (2015b). Liberal and licensed professions. In E. Jones & G. Pasquino (Eds.), *The Oxford handbook of Italian politics* (pp. 541–553). Oxford: Oxford University Press.

Carboni, C. (2015c). *L'implosione delle élite. Leader contro in Italia e in Europa.* Soveria Mannelli: Rubbettino.

Carboni, C., et al. (2017). *Il Divario generazionale tra conflitti e solidarietà. Vincoli, norme, opportunità: generazioni al confronto.* Roma: Dialoghi.

Coletto, D., & Bronzini, M. (2018). Street level bureaucracy under pressure: Job insecurity, business logic and challenging users. In F. Sowa, R. Staples, & S. Zapfel (Eds.), *The transformation of work in welfare state organizations* (pp. 182–202). Routledge: New Public Management and the Institutional Diffusion of Ideas.

David, P. (1989). L'integrazione sociosanitaria: storia di un principio mai realizzato. *Stato e Mercato, 25*(1), 73–109.

David, P., & Vicarelli, G. (1983). *L'azienda famiglia. Una società a responsabilità limitata.* Bari: Laterza.

David, P., & Vicarelli, G. (1994). *Donne nelle professioni degli uomini.* Milano: Angeli.

Giarelli, G., Nigris, D., & Spina, E. (2012). *La sfida dell'auto-mutuo aiuto. Associazionismo di cittadinanza e sistema sociosanitario.* Roma: Carocci.

Ingrosso, M., & Vicarelli, G. (2015). Nascita ed evoluzione della sociologia della salute italiana. In M. Ingrosso (Ed.), *La salute per tutti. Un'indagine sulle origini della sociologia della salute in Italia* (pp. 13–43). Milano: Angeli.

Moretti, C. (2008). L'associazionismo familiare nel sistema di welfare regionale. In A. Genova & F. Palazzo (Eds.), *Il welfare marchigiano: attori, strumenti, criticità* (pp. 235–245). Roma: Carocci.

Moretti, C. (2011). L'integrazione del servizio sociale negli studi medici. In M. Bronzini (Ed.), *Dieci anni di welfare territoriale: pratiche di integrazione socio-sanitaria* (pp. 219–229). Napoli: ESI.

Moretti, C. (2015). La mediazione sociale abitativa nei contesti di edilizia pubblica. In E. Appetecchia (Ed.), *Idee e movimenti comunitari. Servizio sociale di comunità in Italia nel secondo dopoguerra* (pp. 281–296). Roma: Viella.

Moretti, C. (2016). Servizio sociale e salute. In A. Campanini (Ed.), *Gli ambiti di intervento del servizio sociale* (pp. 241–258). Roma: Carocci.

Moretti, C. (2017a). From the hospital towards social reintegration: the support path for people with severe acquired brain injury and their families. *European Journal of Social Work, 20*(6), 858–868. https://doi.org/10.1080/13691457.2017.1320529.

Moretti, C. (2017b). Politiche abitative pubbliche e welfare locale: nuove sfide per il servizio sociale. *RPS, 15*(1), 109–120.

Moretti, C. (2017c). Social housing mediation: Education path for social workers. *European Journal of Social Work, 20*(3), 429–440. https://doi.org/10.1080/13691457.2017.1314934.

Moretti, C., Spina, E., & Ciaschini, U. (2012). Formazione e operatività nel sociale: l'assistente sociale, l'educatore e l'operatore socio-sanitario. *Rivista Trimestrale di Scienza dell'Amministrazione, 3,* 53–72. https://doi.org/10.3280/SA2012-003004.

Moretti, E., & Vicarelli, G. (1986). *I lavoratori stranieri nelle Marche.* Bari: Cacucci.

Novelli, R. (Ed.). (1990). *Industrialization without development: NGOs and growth in South East Asia.* Ancona: Clua.

Orazi, F. (2013). Innovazione sociale e sviluppo locale: i Distretti di Economia Solidale. *Rivista Trimestrale di Scienza dell'Amministrazione, 2,* 63–79. https://doi.org/10.3280/SA2013-002005.

Orazi, F. (2017). False myths and labour market reforms in Italy. *International Journal of Social Science and Technology, 2*(5), 1–11.

Orazi, F. (2018). *Dalla notte dei tempi ai giorni senza tempo.* Milano: Meltemi.

Orazi, F., & Socci, M. (2014). *Il grillismo: tra democrazia elettronica e movimento personale.* Roma: Carocci.

Paci, M. (Ed.). (1980). *Famiglia e mercato del lavoro in un'economia periferica.* Milano: Angeli.

Paci, M. (1982). Onde lunghe nello sviluppo dei sistemi di welfare. *Stato e Mercato, 2*(6), 345–400.

Paci, M. (1989). *Pubblico e privato nei moderni sistemi di welfare.* Napoli: Liguori.

Paci, M. (1991). Classi sociali e società post-industriale in Italia. *Stato e Mercato, 32*(2), 199–217.

Paci, M. (1992). *Il mutamento della struttura sociale italiana.* Bologna: Il Mulino.

Pattarin, E. (2005). La riforma della scuola in Italia. In G. Vicarelli (Ed.), *Il Malessere del Welfare* (pp. 51–70). Napoli: Liguori.

Pattarin, E. (Ed.). (2009). *Traduttori di culture: i mediatori linguistico culturali.* Ancona: Affinità Elettive.

Pattarin, E. (2011). Le politiche scolastiche. In U. Ascoli (Ed.), *Il Welfare in Italia* (pp. 173–195). Bologna: Il Mulino.

Pavolini, E., & Ascoli, U. (2019). The Dark Side of the Moon: il ruolo del welfare fiscale nel sistema di protezione sociale italiano. *Politiche Sociali, 6*(1), 23–46.

Pavolini, E., Ascoli, U., & Mirabile, M. L. (2013). *Tempi moderni. Il welfare nelle aziende in Italia.* Bologna: Il Mulino.

Pizzorno, A. (1974). I ceti medi nel meccanismo del consenso. In F. L. Cavazza & S. R. Graubard (Eds.), *Il caso italiano* (pp. 315–338). Milano: Garzanti.

Speranza, L., Tousijn, W., & Vicarelli, G. (2008). *I medici in Italia, motivazioni, autonomia, appartenenza.* Bologna: Il Mulino.

Spina, E. (2009). *Ostetriche e Midwives. Spazi di autonomia e identità corporativa.* Milano: Angeli.

Spina, E. (2017). Generational gap e nuovo professionalismo medico. Note preliminari per una ricerca empirica. *Rassegna Italiana di Sociologia, 58*(1), 127–152. https://doi.org/10.1423/86361.

Spina, E., & Vicarelli, G. (2015). Are young female doctors breaking through the glass ceiling in Italy? *Cambio, 5*(9), 121–134. https://doi.org/10.1400/234060.

Vicarelli, G. (1986). Professioni e welfare state: i medici generici nel Servizio sanitario nazionale. *Stato e Mercato, 16*(1), 93–122.

Vicarelli, G. (1988). Il personale della salute. In AIS, ISTAT (Eds.), *Immagini della società italiana* (pp. 353–378). Roma: ISTAT.

Vicarelli, G. (1989). Il medico al femminile. Le donne nello sviluppo della professione medica in Italia. *Polis, 2,* 225–248.

Vicarelli, G. (Ed.). (1994). *Le mani invisibili. La vita e il lavoro delle donne immigrate.* Roma: Ediesse.

Vicarelli, G. (1997a). *Alle radici della politica sanitaria in Italia. Società e salute da Crispi al fascismo*. Bologna: Il Mulino.

Vicarelli, G. (1997b). La politica sanitaria tra continuità e innovazione. In F. Barbagallo (Ed.), *Storia dell'Italia repubblicana, Volume III, L'Italia nella crisi mondiale. L'ultimo ventennio* (pp. 569–619). Torino: Einaudi.

Vicarelli, G. (2000). Fiducia e stima nei riguardi della professione medica in Italia. Un'interpretazione di carattere storico-comparativo. *Rassegna Italiana di Sociologia, 41*(3), 389–412. https://doi.org/10.1423/2545.

Vicarelli, G. (Ed.). (2001). *Eliot Freidson: la dominanza medica. Le basi sociali della malattia e delle istituzioni sanitarie*. Milano: Angeli.

Vicarelli, G. (2003). Identità e percorsi professionali delle donne medico in Italia. *Polis, 17*(1), 93–124.

Vicarelli, G. (Ed.). (2005). *Il malessere del Welfare*. Napoli: Liguori.

Vicarelli, G. (2006). Medicus omnium. La costruzione professionale del Medico di medicina generale (1945-2005). In C. Cipolla, C. Corposanto, & W. Tousijn (Eds.), *I medici di medicina generale in Italia* (pp. 50–99). Angeli: Milano.

Vicarelli, G. (Ed.). (2007). *Donne e professioni nell'Italia del Novecento*. Bologna: Il Mulino.

Vicarelli, G. (2008). *Donne di medicina. Il percorso professionale delle donne medico in Italia*. Bologna: Il Mulino.

Vicarelli, G. (2010a). *Gli eredi di Esculapio. Medici e politiche sanitarie nell'Italia unita*. Roma: Carocci.

Vicarelli, G. (2010b). Per un'analisi storico-comparata della professione medica. *Stato e Mercato, 90*(3), 395–424. https://doi.org/10.1425/33148.

Vicarelli, G. (Ed.). (2011). *Regolazione e governance nei sistemi sanitari europei*. Bologna: Il Mulino.

Vicarelli, G. (Ed.). (2013). *Cura e salute. Prospettive sociologiche*. Roma: Carocci.

Vicarelli, G. (2015). Healthcare: Difficult paths of reform. In U. Ascoli & E. Pavolini (Eds.), *The Italian welfare state in a European perspective* (pp. 157–178). Bristol: Policy Press.

Vicarelli, G. (Ed.). (2016). *Oltre il coinvolgimento. L'attivazione del cittadino nelle nuove configurazioni di benessere*. Bologna: Il Mulino.

Vicarelli, G., & Bronzini, M. (2009). From the "expert patient" to "expert family". A feasibility study on family learning for people with long-term conditions in Italy. *Health Sociology Review, 18*(2), 182–193. https://doi.org/10.5172/hesr.18.2.182.

Vicarelli, G., & Bronzini, M. (2018a). Les femmes médecins en Italie: peu de traces, aucune mémoire et une histoire difficile. In G. Ferréol (Ed.), *Traces et mémoires* (pp. 241–250). Louvain-la-Neuve: Eme editions.

Vicarelli, G., & Bronzini, M. (2018b). La sanità digitale: dimensioni di analisi e prospettive di ricerca. *Politiche Sociali, 5*(2), 147–161. https://doi.org/10.7389/90591.

Vicarelli, G., & Pavolini, E. (2017). Dynamics between doctors and managers in the Italian national health care system. *Sociology of Health & Illness, 39*(8), 1381–1397. https://doi.org/10.1111/1467-9566.12592.

Vicarelli, G., & Spina, E. (2015). Professionalization and professionalism: The case of Italian dentistry. *Professions Professionalism, 5*(3). https://doi.org/10.7577/pp.1324.

Money, Banking and Territories: A MoFiR View

Pietro Alessandrini, Luca Papi and Alberto Zazzaro

Abstract This chapter relates the experience of the Money and Finance Research Group (MoFiR) established almost two decades ago within the Department of Economics of UNIVPM. The MoFiR main research focus has been the analysis of the evolution of the financial system on the development of economic systems. This chapter synthetizes some of the authors' main contributions on three main lines of research: the international and European monetary systems, the relationship between banking structures and local development, and the potential discriminatory effects of regulatory policies. In all these issues the main point of reference has been the impact of the evolution of financial structure and regulation on different territories (countries, regions, local systems) characterized by different levels of development. The authors argue that this point of interest should be maintained in any future study on the transformations of the financial system, included the evolving situations of different territories and bank-firm relationships. What has to be always considered crucial is the understanding of how structural changes generate and distribute benefits, risks, and losses within and among economies.

P. Alessandrini (✉) · L. Papi
Department of Economics and Social Sciences, Università Politecnica delle Marche, Piazzale
Martelli 8, Ancona, Italy
e-mail: alepiero3@gmail.com

L. Papi
e-mail: l.papi@univpm.it

A. Zazzaro
Department of Economics and Statistics, Università di Napoli Federico II, Via Cintia 21, Napoli,
Italy
e-mail: Alberto.Zazzaro@unina.it

© Springer Nature Switzerland AG 2019
S. Longhi et al. (eds.), *The First Outstanding 50 Years
of "Università Politecnica delle Marche"*,
https://doi.org/10.1007/978-3-030-33879-4_5

1   Introduction

After 25 years of research in the field of monetary economics in the Department of Economics and Social Sciences of the UNIVPM, in November 2003 Pietro Alessandrini, Alberto Niccoli, Luca Papi and Alberto Zazzaro promoted the institution of Laboratorio Banche, Imprese e Sviluppo (La.B.I.S.) with the aim of favoring a network open to researchers that operated both inside and outside the Department. The excellent results obtained in the first years encouraged the extension of the network at the international level. Hence in November 2007 they denominated the laboratory Money and Finance Research group (Mo.Fi.R.), including Michele Fratianni among the promoting members. The research focus of the group is to investigate, from both the empirical and theoretical points of view, the evolution of the financial system as the collection of financial institutions, intermediaries and markets and the real consequences for the development of economic systems. The main point of reference of our analysis is the impact of the financial structure on territories. The basic view is that the territorial playing field is not levelled. Consequently, also in presence of globalizing technologies and standardized instruments, the functioning of financial intermediaries and markets and their regulatory policies must be reasonably flexible to be adapted to territorial asymmetries. In this way the financial system could better contribute to the reduction of development gaps at regional and international levels.

From 2008 and 2018, the MoFiR group produced a relevant amount of activities with the participation of 29 members and the collaboration of several researchers from other Italian and international universities and public and private institutions. The result has been a wide consideration in the international scientific community. It is remarkable to reckon, among the other initiatives, 150 MoFiR Working Papers and 11 international conferences, of which 8 MoFiR Workshop on Banking.[1]

In the following sections of this chapter we will synthetize our contributions to three main lines of research: the international and European monetary systems, the relationship between banking structures and development, and the regulatory policy.

2   The International and European Monetary Systems

The line of research of the MoFiR group on international monetary issues has been put forward mainly by Pietro Alessandrini and Michele Fratianni. Their interest is based on the consideration of the fundamental disequilibria that persist in the international monetary world. The structural changes, aggravated from 2007 by the great financial crisis, create almost a unique opportunity to reshape the international monetary system (IMS). The big risk is a confidence crisis in the dollar, which is still the dominant currency since the end of the second world war. The consequences would be sharp realignments of exchange rates and the resurgence of protectionism

---

[1] Two conferences had been held in Chicago (USA), one in London (UK), one in Kobe (Japan), all the others at the Faculty of Economics Giorgio Fuà in Ancona.

in international trade. The long run deterioration of the US position from the largest creditor to the largest debtor of the world is the consequence of its policy of benign neglect about external imbalances, benefiting from the "exorbitant privilege" of the issue of the key-currency of the world. This privilege is increasingly unacceptable and cannot persist in the long run. As the new big creditor, China is the critical player in bringing about changes. But the Chinese yuan is far from becoming the new key-currency of the international system. Nor is the euro ready yet to fully replace the dollar, in presence of the incomplete financial and political integration of the European system.

Alessandrini and Fratianni (2009a, b) advocate the creation of a supernational bank money (SBM) within the institutional setting of an international clearing union (ICU). They take inspiration from the five principles underlying Keynes' plan presented at the Bretton Woods Conference in 1943: gradualism, complementarity, multilateralism, banking approach, and symmetry of adjustment. One world money governed by a world central bank is utopian and also difficult to justify in economic terms. In coexistence with national currencies that retain their means-of-payment function, SBMs are owned by existing central banks as multilateral reserve currency vis-a-vis the ICU. Alessandrini and Fratianni's project envisages an initial cooperative agreement among a restricted group of key countries, such us USA, EU, and China, that find in their interest to share responsibility to stabilize the IMS. Hence the ICU should be initially established at the International Monetary Fund (IMF) by three leading central banks: FED, ECB and People's Bank of China. As the clearing system starts and shows its benefits, all central banks will find it convenient to adhere to the ICU. There are two channels of creation of new SBM. First, central banks obtain SBM deposits in change of low risk assets (i.e. treasury bills denominated in national currencies). So SBM has the property of a basket currency with the attendant risk diversifying characteristics. Second ICU operates on a banking principle with the possibility to create SBM allowing central banks to get overdraft facilities on their SBM account. It implies a full-fledged agreement by participating central banks on rules of the game, such as: size and duration of overdrafts, designation of countries that would have to bear the burden of external adjustment, and coordination of monetary policies. The IMF is best positioned to monitor and "enforce" these rules. In their papers Alessandrini and Fratianni analyze the benefits of the SBM proposal. In synthesis: strong reduction of the key-currency countries' "exorbitant privilege", endogenous creation of SBM, revitalized function of the IMF for a better coordination of monetary policy objectives, symmetric responsibility of adjustment shared by external deficit and surplus countries. Moreover Alessandrini and Presbitero (2012) emphasize the adoption of the SBM system, as a step forward from the Special Drawing Rights (SDR) system, in the current foreign assistance framework for a better distribution of development finance to relax the external constraints of less-developed countries.

The second line of research of the MoFiR group concerns the fragility of the Euro-area (EA) exposed to world financial crisis. Again, it is a matter of disequilibria between member countries and the connected problems of financing and adjustment. Alessandrini et al. (2014) analyze the connections between external imbalances and sovereign debt crisis within the EA. The prevailing wisdom is that the South of the EA has been fiscally irresponsible and has failed to implement necessary supply-side policies such as liberalizing labor markets and the market for services. The second view attributes part of the responsibility of the euro crisis to external imbalances resulting from inadequate adjustment mechanisms. The North (in particular Germany) has enjoyed large current-account surpluses, while the South has accumulated current account deficits, suggesting that real exchange rates are too weak for the North and too strong for the South. The paper shows that these two views are not inconsistent. The empirical results confirm that the EA sovereign debt crisis has been as much a matter of external imbalances as of fiscal irresponsibility. At the basis of both problems lie the inadequate adjustments in competitiveness between surplus and deficit countries. Since nominal exchange rates are fixed in the EA, and given that the level of economic activity is historically low in the South, an expansion of aggregate demand in the North and less austerity in the South would have worked better than strict austerity over the entire EA.

Alessandrini and Fratianni (2015) consider that, in the absence of fiscal union, the correction of external imbalances within the EA must be taken as seriously as that of national fiscal imbalances and debt-to-GDP ratios. The EU Commission should promote set targets on current-account imbalances, applied symmetrically to both deficit and surplus countries. Moreover, the ECB should adopt a managed flexibility of the common monetary policy. On the surface, there could be a contradiction between a common monetary policy and the introduction of some flexibility. The principle of a unified monetary policy is maintained once ECB fixes the same strategy and the same basic official rates of interest for all EA countries. The specific proposal of the paper is that national central banks of the EA should add a risk premium cost to official interest rates on banks that accumulate "excessive" borrowings or deposits to compensate, respectively, for outflows and inflows of the monetary base due to the effect of external imbalances. This solution, aimed to contrast such institutional sterilization, would favor the monetary adjustment of external disequilibria.

3   Geography of Banking Industry, Distances and Development

Since the early nineties, research on banking and financial issues in Ancona has focused on the geographical organization of banking systems and their impact on regional development. This was an innovative line of research on which at that time an intense debate was developing at both the national and international level, spurred by the ongoing merger and acquisition processes in many banking systems around the world, and that would then find its place in the most prestigious international journals.

The common opinion among scholars and practitioners was that deregulation, advancements in information technology and financial innovations would have made banking activity ever more transaction oriented, including a great number of non-traditional financial products. A natural consequence of this trend would be the expansion of banks on a global scale both geographically and in terms of products supplied. A supranational financial system would have emerged, populated by few global players able to offer standardized products in many different countries and regions. The importance of geographical proximity between banks and borrowers would be doomed to decline over time and 'the end of financial geography' would become a real possibility (O'Brien 1992).

Some critical voices warned against the myth of global banking, pointing out the role of banks as agents of development, and the high costs that the consolidation and geographical agglomeration of bank decisional centres would generate for peripheral regions and small local firms. Among these are the work of Alessandrini (1989) on regional financial flows, the analyses by Alessandrini (1992, 1994, 1996) and Zazzaro (1998) on the territorial articulation of the banking system in some Italian regions. Alessandrini and Zazzaro (1999) propose a "possibilist" approach taking into account cost and benefits of financial integration for the peripheral regions and indicate the need to favor processes of active integration "that are best suited to the specific and varying characteristics and adjustment capabilities.

This approach was used to warn against the specular myth of local banks. Although the informational advantages of local banks, stemming from their historical roots and "cultural affinities" with the local community, allow for sounder assessment of local firms, they do not ensure that credit is always allocated as best suits local economic development. First, in backward areas, liquidity costs may discourage local banks from financing innovative businesses. Since local banks concentrate most of their deposit collection activity in limited geographical areas, the reflux of deposits resulting from the granting of a loan will be higher, and the cost of liquidity for the bank will be lower, as greater the amount of granted credit spent locally by borrowers as happens in the case of less innovative firms, operating in domestic-demand-oriented industries (Zazzaro 1993, 1997). Second, in-depth exclusive knowledge of a single economic environment may reduce a local bank's capacity to react to novelties from the world of production, and long-term ties with local firms may drive local banks to limit the entry of new firms and the financing of highly innovative businesses. This type of attitude on the part of local banks would also end up reducing the innovative efforts of existing firms which, "protected" by the local banks, would have less incentive to introduce innovations (Zazzaro 2002).

This first strand of studies is ideally completed by the article of Lucchetti et al. (2001). In this study, they propose a novel approach for measuring financial development based on the efficiency of the banking industry, which has been widely used in the literature on finance and development. Namely, they introduced a measure of the inefficiency of regional banking systems based on the microeconomic cost inefficiency of individual banks, weighted for the respective share of bank branches in the region, and included this measure in a panel convergence regression for Italian regions. Their findings indicate that the inefficiency of the regional banking systems has a significant negative influence on regional GDP growth rates independent of the presence of local banks, overall lending to local firms and other possible confounding factors (human capital, efficiency of the local judiciary system, and regional and time dummies).

The processes of privatization, deregulation and consolidation of banking industry in the 90s led to an increasing spatial concentration of bank decision-making centres and strategic functions in a few places within each country. At the same time, however, the geographical diffusion of banking structures and instruments expanded in almost every industrialized country (Alessandrini et al. 2009b). This spatial concentration-diffusion trend in the banking industry raised some new important issues about the role of relationship banking in the era of global banking, the real effects of the organizational structure of a local banking system, and the impact of the geographical and cultural distances between bank operational branches, bank decisional centres and borrowers on loan contracts, lending technologies and credit allocation.

Using the lens of geography of banking power, related to the spatial distribution of credit institutions involved in M&A deals in Italy, Alessandrini et al. (2005) find that, unlike what happened in the Centre-North, acquired banks headquartered in less developed Southern regions show worse performance indicators (small business lending, loan growth, bad loans and profitability) than stand-alone banks located in the same area, regardless of time elapsing from acquisition and the size of the acquired bank. Alessandrini et al. (2008) focus on the portfolio restructuring strategies of banks involved M&A deals. They show that acquisitions involving banks headquartered in Central and Northern Italy were dominated by a simple asset-cleaning strategy by the bidder bank, without changing the asset allocation of the target bank. By contrast, in the case of acquisitions of banks in the Southern regions by banks of the Centre-North, the asset restructuring strategy pursued by the acquiring bank led to a structural change in the portfolio of the acquired bank with a permanent reduction in loans to small firms and an increase in asset management activity, which was more pronounced as the cultural distance (in terms of social capital) between the provinces where dealing partners are headquartered was greater.

The changing geography of banking industry was the subject of the first international conference organized by the MoFiR group in September 2006. On that occasion, the notion of "functional distance", introduced by Alessandrini et al. (2005), was for the first time extended to local banking systems as a whole by Alessandrini et al. (2009a) who operationalized it by the index:

$$FD_j = \frac{\sum_{b=1}^{B_j} [branches_b * \ln(1 + D_{jzb})]}{\sum_{b=1}^{B_j} branches_b}$$

where B_j is the number of banks operating in province j, $branches_b$ is the number of branches belonging to bank b in the province j, and D_{jzb} is the geographical or cultural distance to the province j where the bank b is headquartered.

In a number of contributions, it was shown that in Italy firms located in provinces disproportionally populated by functionally distant banks have less access to credit

(Alessandrini et al. 2009a), a lower capacity to maintain a long-lasting bank relationship (Presbitero and Zazzaro 2011), a lower propensity to innovate (Alessandrini et al. 2009b), and suffered from a stronger credit crunch during the financial crisis (Presbitero et al. 2014). The negative effects of the functional distance of the local banking system tend also to mitigate the positive effects for firms' access to credit and lending relationships of being located in an industrial district (Alessandrini et al. 2008; Alessandrini and Zazzaro 2009).

Finally, Bellucci et al. (2013, 2019) and Filomeni et al. (2016) analyze the role of operational and functional distance on loan contract terms and information production.

4 The Effects of the Regulatory Process

In recent years the attention of the MoFiR group has extended to the analysis of structural aspects of the banking sector due to a new regulatory wave. It is a known fact that new financial regulation comes in big swings. The two decades before the 2007 financial crisis had been characterized as a liberalization period from barriers between commercial banking, investment banking and insurance. The new deregulation established an environment of liberalized capital flows, facilitating the growth of large, complex and interconnected banks. More recently, as a result of the 2007 financial crisis, the deregulation/re-regulation pendulum has shifted again towards the stability of the banking sector, bringing with it a new wave of rules. Consequently, as the effects of the crisis have settled, two main aspects of a new operative scenario in banking have emerged (Alessandrini et al. 2016a).

The first aspect is the end of the liberalization process that was supposed to improve the efficiency of banks and financial markets. In fact, years of deregulation had enhanced the rise of big universal banks, unanimously considered responsible of the contagion in the crisis through the spread of the originate-and-distribute model. The second one is the introduction of a growing regulatory system (the so-called Basel III) to forestall financial risks and regain banking stability. Under the pressure of the crisis, Basel III emerged as a much stronger version of Basel II in terms of capital requirements. It also introduced non-credit risk-based measures such as ceilings on the leverage level and liquidity ratios. Basel III is actually an ongoing process that produces a constant flow of new norms and clarifying documents. But the one feature that really stands out is its complexity (Masera 2015, Alessandrini and Papi 2018). Complexity goes beyond the enormous number of pages detailing rules and interpretation. It is measured in terms of data, analytics, implementation and reporting requirements. Judged in terms of interaction alone, complex controls are at a disadvantage compared to simple controls. But while Basel III defines a forest of risk definitions, it ignores that ultimately risk is determined by the interaction of a complex system with complex controls (Caprio 2013). Furthermore, implementation and compliance costs are another strike against complexity. The effects of this regulatory evolution have been widely studied in relation to many aspects. For instance,

Fratianni et al. (2017) assess the link between regulation and the probability of a banking crisis and show how the probability depends on factors such as the kind of regulation and more interestingly on the quality of institutions.

Less studied by the economic literature and above all less considered by the competent authorities have been the consequences of the increased regulation on the structural characteristics of banking systems. A recent focus of the MoFiR research agenda is how this new regulation is affecting the structure of the banking system and, in particular, which asymmetric effects are emerging. Two kinds of asymmetries have been considered.

The first asymmetry is on bank size. While stability comes with a cost, the working principle is that this cost be distributed proportionately across different types and sizes of banking institutions. A vast and complex regulatory system is bound to alter banks behavior and a uniform regulation may end up having asymmetric effects on different types and sizes of banks. Much of regulation is a fixed cost, which creates a proportionally higher burden on small banks than on large banks. The requirements of new regulation in terms of administrative and IT costs are for all kinds of banks, whereas their impact should take into account criteria such as the size, business model and complexity of banks (Alessandrini et al. 2016b). In contrast, uniform regulation violates the important principle of proportionality which should be recognized and applied to many aspects of the regulatory process. In the presence of an insufficiently applied proportionality principle, a regulatory system of increasing size and complexity can influence the level of competition within the banking system. This effect leads to the raising of barriers to entry into the sector, imposing fixed costs from the heterogeneous effects on the various size categories of banks.[2]

The second asymmetry of the new regulation is on territories. In this respect, we can ask how the regulatory costs are transferred to bank customers. With regard to the credit granted, we can distinguish between a quantity effect and a price effect. The credit supply has certainly been influenced by the regulation that has innovated on the reclassifications of credits and imposed additional provisions.[3] The recognition in the financial statements of the new value adjustments, required by the regulation and due to the deterioration in the quality of credit, affected negatively the supply of loans. On the other hand, the price effect depends on the elasticity of the demand for credit, which in turn is a function of accessibility to cheaper alternative financial solutions. Thus it is possible that those customers that, for example, may resort to alternative forms of financing are less affected by higher banking costs, while on the contrary, households and small businesses, for which there are practically no alternatives to banking products, are probably forced to bear a greater burden. If we then consider the specialization of small banks, which focuses on SMEs and peripheral territories, we can conclude that regulatory responses to the banking and

---

[2] See Alessandrini and Papi (2018) for some empirical evidence of the different regulatory burden for the different size categories of banks for the Italian case.

[3] According to the Bank of Italy, an increase of one percentage point of the new non-performing loans in relation to total loans would correspond to an average reduction of the growth rate of loans to companies by about 1.5 percentage points (Bank of Italy 2017).

financial crisis weigh more on small businesses and those peripheral territories that will continue to depend on small local banks. Moreover, in less developed areas, banks face a higher proportion of riskier firms. Since these firms impose a higher consumption of bank capital, uniformly stricter capital-based rules are bound to amplify regional disparities.

In fact despite these clear asymmetries, Basel III treats all banks virtually the same.[4] This uniformity affects unfavorably the small local or community banks that are an important component of many banking systems, including the United States and several European countries, Italy first. Moreover, large banks receive a "too big to fail" subsidy from the possibility that the capital surcharge may not be adequate to prevent being rescued by governments. Without a regulatory correction, small banks are at risk of disappearing, an issue that is hotly debated in the United States, which has already implemented a dual-regulatory system, one applicable to very large banks (Advanced Approaches Banks) and another to community banks (Fratianni 2015). The latter face smaller risk-weighted capital ratios than the former and are exempt from the countercyclical capital buffer, supplementary leverage ratio, and credit valuation adjustments requirements. Furthermore, community banks in the United States are subject to lighter supervision than applicable to large banks and are exempt from stress testing and capital planning requirements (Yellen 2014). In contrast, the EU application of Basel III does not make any substantial distinction between large and small banks. In fact, with the exception of the global systematically important banks, the European regulatory approach envisages a sort of "one-size-fits-all" regulation framework relegating the implementation of the principle of proportionality basically to a different frequency of the supervisory engagement for the various size categories of banks

To conclude, regulation impacts not only risk and profitability of the banking system as a whole but also its structure. The last wave of regulation is relatively unfriendly to local banks, reflecting the position of regulators, especially European, that a consolidation of the banking system can lower systemic risk. Our belief is that it should not be the regulator to select the preferred type of bank. The task of evaluating the efficient structures should be left to the market. In general, the principles of the diversity of the structures and the proportionality of the control system must be objectives to be exploited rather than cancelled, to the benefit of competition that produces efficiency and the diversification of business models that produces stability.

5 Conclusions

The three main lines of research developed by the MoFiR group deal with problems that are still open. However, the results obtained will maintain their value for future researches. Everything is subject to changes. Technological innovations are the main

---

[4]Except the 30 global systematically important banks (G-SIBs).

driver of structural transformations and the financial system is particularly sensitive to innovations in means of payments, financial instruments and services, structures of intermediation, and forms of control and policy. The fintech revolution is already in action.

In this rapid evolution with unforeseeable effects, it is important to maintain a clear point of reference in the method of research. This approach has produced significant results in the researches we have synthesized in this chapter. They refer to the last three decades, that had already been characterized by important transformations, like: the relative decline of the dollar supremacy, the increased role of China as a creditor country, the birth of the Eurosystem, the consolidation process of banking structures, the evolution of the lending relationships with bank customers, the epidemic impact of the great financial crisis, the consequent passage of the regulatory pendulum from a long period of liberalization to a pervasive system of regulation.

In all these problems the focus of our analysis has been the evaluation of their impact on territories (countries, regions, local systems), characterized by differences in levels of development. This point of reference should be maintained for any future transformations. What is always needed is to understand how any structural innovation will generate and distribute benefits, risks, and losses. Moreover, it is important to monitor the potential destabilizing effects of fundamental imbalances generated by excess of financing and difficulties of adjustment. This point of reference always applies in the external exchanges between countries, regions, and in bank-firm lending. And it will remain a fundamental factor of instability whatever future financial system will prevail. Finally, the regulation structure must always be flexible and not discriminatory. Flexibility is needed to obtain a difficult but necessary balance between financial efficiency and stability. The no discriminatory impact is requested to maintain a diversified structure of financial intermediaries functional to the diversified structure of firms.

References

Alessandrini, P. (1989). I flussi finanziari interregionali: interdipendenze funzionali ed indizi empirici sulla realtà italiana. In A. Niccoli (Ed.), *Credito e sviluppo: Evoluzione delle strutture e squilibri territoriali* (pp. 223–279). Giuffrè: Milano.

Alessandrini, P. (1992). Squilibri regionali e dualismo in Italia: alcune riflessioni. *Moneta e Credito, 47,* 67–81.

Alessandrini, P. (Ed.). (1994). *La banca in un sistema locale di piccole e medie imprese.* Bologna: Il Mulino.

Alessandrini, P. (1996). Sistemi locali del credito in regioni a diverso livello di sviluppo. *Moneta e Credito, 50,* 567–600.

Alessandrini, P., & Fratianni, M. (2009a). Resurrecting keynes to stabilize the international monetary system. *Open Economies Review, 20*(3), 339–358.

Alessandrini, P., & Fratianni, M. (2009b). Dominant currencies, special drawing rights, and supranational bank money. *World Economics, 10*(4), 45–67.

Alessandrini, P., & Fratianni, M. (2015). In the absence of fiscal union, the eurozone needs a more flexible monetary policy. *PSL Quarterly Review, 68*(275), 279–296.

Alessandrini, P., & Papi, L. (2018). L'impatto della bolla regolamentare sulle banche (pp. 2–20). Bancaria.

Alessandrini, P., & Presbitero, A. F. (2012). Low-Income Countries and an SDR-based international monetary system. *Open Economies Review, 23*(1), 129–150.

Alessandrini, P., & Zazzaro, A. (1999). A "possibilist" approach to local financial systems and regional development: The Italian experience. In R. Martin (Ed.), *Money and the Space Economy* (pp. 71–92). New York: Wiley.

Alessandrini, P., & Zazzaro, A. (2009). Banking localism, distances and industrial districts. In G. Becattini, M. Bellandi & L. Depropris (Eds.), *A handbook of industrial districts* (pp. 471–482). Cheltenham: Edward Elgar.

Alessandrini, P., Calcagnini, G., & Zazzaro, A. (2008a). Asset restructuring strategies in bank acquisitions: Does distance between dealing partners matter? *Journal of Banking & Finance, 32*(5), 699–713.

Alessandrini, P., Croci, M., & Zazzaro, A. (2005). The geography of banking power: The role of functional distances. *BNL Quarterly Review, 58*(235), 129–167.

Alessandrini, P., Fratianni, M., Hallett, A. H., & Presbitero, A. F. (2014). External imbalances and financial fragility in the euro area. *Open Economies Review, 25*(1), 3–34.

Alessandrini, P., Fratianni, M., Papi, L., & Zazzaro, A. (2016a). Banks, regions and development after the crisis and under the new regulatory system. *Credit and Capital Markets, 4,* 536–561.

Alessandrini, P., Fratianni, M., Papi, L., & Zazzaro, A. (2016b). The asymmetric burden of regulation: Will local banks survive? In G. Bracchi, U. Filotto, & D. Masciandaro (Eds.), *The Italian banks: Which will be the "new normal"? 2016 report on the Italian financial system* Roma: EDIBANK.

Alessandrini, P., Presbitero, A. F., & Zazzaro, A. (2008b). Banche e imprese nei distretti industriali. In A. Zazzaro (Ed.), *I vincoli finanziari alla crescita delle imprese* (pp. 244–266). Carocci: Roma.

Alessandrini, P., Presbitero, A. F., & Zazzaro, A. (2009a). Banks, distances and firms' financing constraints. *Review of Finance, 13*(2), 261–307.

Alessandrini, P., Fratianni, M., & Zazzaro, A. (2009c). Geographical organization of banking systems and innovation diffusion. In P. Alessandrini, M. Fratianni & A. Zazzaro (eds.), The changing geography of banking and finance. New York: Springer, 75–108.

Alessandrini, P., Presbitero, A., & Zazzaro, A. (2010). Bank size or distance: What hampers innovation adoption by SMEs? *Journal of Economic Geography, 10*(6), 845–881.

Bank of Italy. (2017). *Report on financial stability*, November.

Bellucci, A., Borisov, A., & Zazzaro, A. (2013). Do banks price discriminate spatially? Evidence from small business lending in local credit markets. *Journal of Banking & Finance, 37*(11), 4183–4197.

Bellucci, A., Borisov, A,, Giombini, G., & Zazzaro, A. (2019). *Collateralization and distance*. Journal of Banking & Finance, 100, 205–217.

Caprio, G., Jr. (2013). *Financial regulation after the crisis: How did we get here, and how do we get out?* LSE financial market group special paper series N. 226, November.

Filomeni, S., Udell. G. F., & Zazzaro, A. (2016). *Hardening soft information: How far as the technology taken us?* MoFiR working paper no. 121.

Fratianni, M. (2015). Basel III in reality. *Journal of Economic Integration, 30*(1), 1–28.

Fratianni, M., Pisicoli, B., & Marchionne, F. (2017). *Regulation, financial crises, and liberalization traps*. MoFiR working paper no.143.

Lucchetti, R., Papi, L., & Zazzaro, A. (2001). Banks' inefficiency and economic growth: A micro-macro approach. *Scottish Journal of Political Economy, 48*(4), 400–424.

Masera, R. (2015). *Regole e supervisione delle banche: Approccio unitari vs modello per livelli e implicazioni per la morfologia del sistema delle banche, EU e US*. Working paper, Università degli Studi Guglielmo Marconi.

O'Brein, R. (1992). Global financial integration: The end of geography. Royal Institute of International Affairs. London.

Presbitero, A. F., & Zazzaro, A. (2011). Competition and relationship lending: Friends or foes? *Journal of Financial Intermediation, 20*(3), 387–413.

Presbitero, A. F., Udell, G. F., & Zazzaro, A. (2014). The home bias and the credit crunch: A regional perspective. *Journal of Money Credit and Banking, 46*(s1), 53–85.

Yellen, J. L. (2014). Tailored supervision of community banks, remarks made at policy summit of the independent community bankers of America, Washington, May 1.

Zazzaro, A. (1993). Banche locali e sviluppo economico regionale: Costi di liquidità e costi di solvibilità. *Rivista di Politica Economica, 107*–152.

Zazzaro, A. (1997). Regional banking system, credit allocation and regional economic development. *Economie Appliquée, 51*(1), 51–74.

Zazzaro, A. (1998). Articolazione territoriale del sistema bancario e sviluppo economico: Aspetti teorici ed alcune evidenze relative alla Campania. *Moneta e Credito, 51*(203), 295–330.

Zazzaro, A. (2002). The allocation of entrepreneurial talent under imperfect lending decisions. *Rivista Italiana degli Economisti—The Journal of The Italian Economic Association, 7*(3), 303–330.

Destination Europe: The Transformation of Agriculture Between Decline and Renaissance

Franco Sotte, Roberto Esposti and Silvia Coderoni

Abstract The objective of the work is to retrace the main contributions of 50 years of studies and research by the group of agricultural economists of the Faculty of Economics of the University of Ancona/Università Politecnica delle Marche. These contributions share a common vision: the transformations of the primary sector in Italy and in the Marche region, are the original result of two equally powerful but largely independent forces. On the one hand, a tumultuous manufacturing development, territorially unbalanced and passively suffered. On the other hand, the centrality assigned to the agricultural sector by a Europe under construction and which has made the primary sector a privileged laboratory, not only as a recipient of resources, but also as an outpost in the definition of policy strategies and design. The originality of the contributions of this group of scholars lies in the analysis of these transformations from an openly critical perspective towards the then prevalent, often unilateral and hagiographic, readings of a development model and a certain idea of Europe. That model is now at an end; and that idea of Europe must be overcome. It is therefore this critical perspective that makes these contributions still relevant and the driving force of present and future research.

F. Sotte · R. Esposti (✉) · S. Coderoni
Department of Economics and Social Sciences, Università Politecnica delle Marche, Piazzale Martelli 8, 60121 Ancona, Italy
e-mail: r.esposti@univpm.it

F. Sotte
e-mail: f.sotte@univpm.it

S. Coderoni
e-mail: s.coderoni@univpm.it

© Springer Nature Switzerland AG 2019
S. Longhi et al. (eds.), *The First Outstanding 50 Years of "Università Politecnica delle Marche"*,
https://doi.org/10.1007/978-3-030-33879-4_6

71

1 Introduction: On the Rhetoric of the "Widespread Development"

This chapter aims to summarize the main contributions provided by the agricultural economists of the "Ancona school" (i.e. Ancona Agricultural Economists, AAE henceforth) over the last 50 years, as well as the consequent main challenges for their future research activity. This group and different generations of scholars represented a sort of heterodoxy within the Ancona school in the way they investigated the Italian (and, then, European) regional development processes from a peculiar perspective, that of a "losing" sector, agriculture, and of "losing" territories, the rural areas. From this perspective, the major limits and unsustainable features of those development processes became progressively manifest and were analysed and pointed out in many different works, either journal articles or books.

The Faculty of Economics (and in particular its current Department of Economics and Social Sciences) of the Ancona University (now Università Politecnica delle Marche) has always been recognized among the national and international community of economists for one major reason: its tradition (the "Ancona school") on studying the long-term development process of that part of Italy (the North-East-Centre-NEC or the "Third Italy") based on localised systems of specialised small and medium enterprises (the "industrial districts"). Hundreds of books and papers have been written on this apparently successful and peculiar "way of development" and, for this reason, often designated as the "NEC model". Not only has the school's contribution been considered relevant but also significantly heterodox with respect to mainstream development economics allegedly incapable of seeing the specificity of this "model" and of understanding its foundations.

The contribution of the "Ancona school" to the identification and definition of the NEC model has been remarkable. After all, in the Faculty website this tradition and this "mission" is still very explicitly declared. It is written that "in their respective fields of interest […] the Ancona scholars have always endeavoured to analyse the problems of the Italian economy and society […]. The characteristics of Italy's long-term development, the dualism of its labour market, the emergence of the Italian industrial districts and their relation with the civic values of the north-east and central regions of the country are among the achievements that have contributed to the recognition of the Faculty's research within and outside Italy". It is also worth remembering that the Faculty is named after Giorgio Fuà, maybe its most famous and valuable scholar. On the Fuà legacy, the Faculty website declares that: "Fuà studied the economy of Italy and especially of Marche; he was the first to interpret the Marche model of development and grasp its structural characteristics: a myriad of small businesses rooted in the territory". Eventually, "Fuà was the inspiration behind and interpreter of what in later years was to become known as the 'NEC model'" (Fuà and Zacchia 1984).

But, then, came the post-2008 great economic crisis. Allegedly unpredicted and unexpected by most researchers, the economist community was suddenly convinced that the crisis was a clear demonstration of the failure of mainstream economics, the same mainstream that did not even see the successful "NEC model". Unfortunately, the crisis dramatically struck also the NEC economy and society, and the Marche region in particular. After all, the global economy quite rapidly and successfully recovered from the post-2008 crisis. The Marche economy, and large parts of the NEC economy, never did. Therefore, it seems now intellectually honest to replace the word "crisis" with "long-term decline". Even though it may seem bizarre, those same economists that celebrated the success of the "model" and have been very prompt in celebrating the funeral of mainstream economics after the crisis, never celebrated the funeral of the "model". Apparently, because they did not even see its death or, maybe, because this could implicitly imply also the death of that whole tradition.

The real question is whether and how the Ancona school really contributed to the understanding of its own regional growth process and its apparently unstoppable decline. In fact, there was a heterodoxy within the school of alleged heterodox. This other "heterodoxy" actually came from those that over years criticised the emphasis on the NEC/Marche model suggesting its possible long-term fallacies: environmental degradation, excessive geographical concentration with consequent congestion, loss of sectors, inadequate integration with the European and global economy. Their main intuition was that such development process was not eventually pushing this part of Italy closer to the core of Europe: it was not becoming the Southern limit of the European core but just the Northern limit of its periphery.

Several works insisted on the agricultural foundations of that development experience (Bartola 1979, 1983a; Esposti and Pierani 1995; Sotte 1996; Esposti and Sotte 2011; Esposti 2012a).[1] The NEC model originated in regions that were still distinctly agricultural (in the Marche region in 1951 agriculture accounted for 60.2% of the total employment) and where the sharecropping contracts were widely adopted and often prevalent. By investigating the salient features of the sharecropping contract, and the underlying decision-making mechanisms in particular, these studies pointed out how the routine of managing a complex production organization with the consequent resource mobility made sharecropping the basis for the genesis and success of the region's manufacturing business network and made the tenant the prototype of that entrepreneurial attitude.

According to this analysis, the determinants of the NEC model were rooted in its agricultural origin. It represented the key endogenous force of that peculiar regional development experience. A first main implication of this conclusion was that such "model" could not be easily replicated in other territories. So, maybe, it was not a "model" at all. A second major implication concerned the long-term sustainability of that development process. As there were the peculiarities of agriculture at the foundation of the entrepreneurial vitality of that economy, a lasting development could not be possible if not maintaining these peculiarities and, in particular, the integration between sectors (agriculture and manufacturing) and territories (the rural and the urban space).

---

[1] For other contributions on this aspect see also Anselmi (1978, 2000).

In fact, the excitement, and a bit of propaganda, raised by the manufacturing success was so strong in those years that, instead of understanding the complexity of their relations with the territory and rural society, farmers themselves were asked to adapt their production processes and decision making to the industrial logic. Eventually, the NEC model forced agriculture to "industrialize" itself. This conclusion was at the heart of a second major original contribution of this group of agricultural economists. They provided clear evidence that the highly celebrated development experience was in fact destroying that integration and was, on the contrary, generating an internal dualism. This dualism was not only jeopardizing the long-term sustainability of that model, but it was also opening many critical, and more general, questions on the role of agriculture and of the rural space in the future of industrial and post-industrial economies and societies (Sotte 2008, 2013; Esposti 2012b, 2014).

2 On the Dark Side of the "Model"

Several studies and research projects particularly in the eighties and nineties revealed that the allegedly balanced development process was in fact substantially unbalanced and dualistic across sectors and territories (Bartola 1983b; Bartola et al. 1984; Esposti and Sotte 2002b; Esposti 2004). In the end, when seen from the perspective of the "losing" sectors and areas there was no real difference compared to the industrialization experience of the rest of Northern Italy and, more generally, of most of the western industrialised countries. Two aspects of this polarization, in particular, deserved attention: the decline of agriculture, the marginalization of rural areas.

While providing a fundamental contribution to the rise of that industrialization experience, the development process itself induced a rapid decline of agricultural activity and entrepreneurship. Attracted by the intense and localised growth of manufacturing, farming was progressively emptied of workforce (especially the younger and the more educated one), of capital and of entrepreneurial capacity. This led to an oversimplification of the agricultural production systems originally based on the strong complementarity between cattle breeding and cultivation that had also favored land fertilization, soil conservation and water regulation (Sotte 1987; Arzeni et al. 2001; Coderoni and Esposti 2014, 2018; Baldoni et al. 2018).

The primary sector passively adjusted to the general socio-economic evolution by acquiring, to remain viable and profitable, hyper-simplified and speculative characters: an "urban and industrial" agriculture that was functional to the areas of strongest agglomeration. The loss of competitiveness of such an "impoverished" sector inevitably weakened the entire agri-food industry (Esposti and Listorti 2009). The natural, physical and social capital accumulated within these farming systems over centuries was quickly sacrificed to make the "miracle" of widespread industrialization take place. But that "miracle" left agriculture almost without a long-term perspective, dependent on policy support, incapable of competiveness on the global markets. A sort of vestigial activity.

But the dark side of the model was even more evident when looking at the territorial dualism it generated. One of the main contributions of the AAE was to emphasize that the widespread character of that development process actually resided in its initial spatial settlement. It was the centuries-old accumulation process, that eventually activated the rapid industrialization process, to be widespread, that is, diffused over space. In fact, the industrialization dynamics was strongly centripetal with a major concentration of resources, from people to infrastructure, in a limited urban portion of these regions. Therefore, it could be legitimate to refer to "rural industrialization" with the exclusive meaning of rural origin not of an actual involvement of the more rural territories, for instance the internal areas of the Apennines. On the contrary, these latter were substantially excluded by that convulsive industrialization and accumulation process.

This original view on the dualistic nature of the "model" led this group of scholars to less hagiographically envisage its long-term perspective and this eventually turned out to be somehow prophetic. According to this evolutionary interpretation that process could not be sustained simply because it was taking advantage of a secular accumulation that was itself leading to a quick end. The strength of its genuine accumulation process was, in fact, secondary compared to the transfer of capital accumulated over centuries in its overlooked agricultural and rural space. Meanwhile, the inevitable slowdown, and then extinction, of the industrialization process was expected to increasingly reveal its drawbacks, the growing consequences of its unbalance: environmental degradation, urban congestion, increasing inequality, social conflicts, loss of economic and social resilience (Esposti and Sotte 1999, 2000). Paradoxically, while the successful development process had to be regarded as transitory, its drawbacks could become permanent features of these economies and societies.

It is worth noticing that the risk of extinction of that development experience has been envisaged by several scholars since the seventies. In these cases, however, the main argument was the progressive disappearance of a series of advantageous external conditions and, therefore, the loss of competiveness in the new global landscape. The original contribution of the AAE, and of a few others (Calafati 2008), in fact, insisted on the internal forces of the model and, therefore, on the fact that extinction rather came from the exhaustion of its internal drivers. Essentially, it was self-extinction.

3   On the Renaissance of Rurality

On this rural perspective of the regional development processes, the nineties opened new and somehow unexpected research lines on which the AAE concentrated most of their attention over the decade. These topics can be grouped under the name of "rural renaissance" and express the new interest on the rural dimension of the post-industrial development, thus also on the role of agriculture in this respect. As a matter of fact, this new interest did not originate within the scientific and intellectual debate on the Italian regional development. It actually came from the international context and, in particular, from the attention paid by European Commission, OECD, World Bank on the peculiar development trajectories of rural territories especially when they did not experience any significant involvement in major industrialization processes (Esposti and Sotte 2001b; OECD 2006).

The research challenge for the Ancona scholars was not limited to understanding whether and how those rural local systems were excluded from the abovementioned accumulation and agglomeration processes. There was something much more general. The key research question concerned the future of post-industrial economies and societies and the role of agriculture and of the rural space within the post-fordist organization of production and consumption. The exclusion from an intense, and somehow disruptive, industrialization process could become a strategic asset for many territories. Rurality was no more a synonymous of backwardness and marginalization. The increasing demand for higher quality products as well as for higher living standards, also from an environmental perspective, together with the emergence of new technologies strongly attenuated the burden of distances and small scale, making many rural territories increasingly interesting and even desirable for businesses and professions which were once exclusively focused on large urban centers (Esposti and Berloni 2001; Esposti and Sotte 2001a, 2002a).

These new opportunities could be seized particularly by those sectors with prevalent "rural" features, agriculture in the first place. A new generation of farmers, female farmers included, and a new a generation of farms diversifying the activity across different agricultural and non-agricultural businesses (the so-called multifunctional agriculture), started to emerge (Sotte 1997, 2006; Finocchio and Sotte 2006; Arzeni et al. 2014). To this new agriculture and rural economy were assigned new social functions sometime summarized under the term "stewardship", that is, ensuring, on the behalf and in the interest of the local and global community, the adequate provision of vital public goods and services, from climate change mitigation and food safety to landscape and conservation of cultural traditions.

Like the process it aims to investigate, this research agenda is still in progress looking for regularities, determinants and appropriate metrics. Nonetheless, the main merit of the AAE, in this respect, was to envisage the new post-industrial and post-fordist pivotal role of the agricultural and rural space starting from the NEC experience. This new role could mitigate and even revert those polarizing forces and their consequent dualism eventually making the "diffusion" not an original character of the development process but, in fact, its desirable outcome. Ancona scholars not only investigated the main features of this rural and agricultural renaissance and the critical role of non-conventional factors (research, knowledge, education) in this respect (Esposti 2000, 2002, 2004, 2011a; Esposti 2012c; Esposti and Pierani 2003; Esposti and Materia 2016); they also identified the conditions to make this renaissance viable also for those local rural contexts so strongly depleted by the allegedly "widespread industrialization". For this new generation of farms and farmers to emerge within these impoverished local contexts, an active role of the "state" was needed through the design and the implementation of appropriate targeted policies.

For this main reason, most of the research activity of the last two decades of the AAE actually concentrated on this renewed role of the "state" and on the analysis of the consequent agricultural, rural and regional policies. It was clear, in particular, that dealing with agriculture and the rural space, the "state" was actually the European Union and these policies were most exclusively the EU policies (Buckwell and Sotte 1997; Esposti and Bussoletti 2008).

4   On Why Europe Matters

From their peculiar perspective, the AAE pointed out the centrality of the EU and of the respective policies when many of their colleagues still barely recognized the very existence of this European dimension. This merit actually depends on the fact that until the late eighties, the Common Agricultural Policy (CAP) has been the largely prevalent, if not the only, EU policy. In fact, the CAP is still granted about 40% of the EU budget. Therefore, already in the seventies, the AAE could apply their long-term experience in the analysis and assessment of national and regional agricultural policies (Orlando 1965; Bartola and Sotte 1983, Bartola and Sgroi 1984) also to this new dimension that had to become soon the only real policy for the agricultural and rural space in Italy.

The contribution of these Ancona scholars to the analysis of the CAP is remarkable and internationally acknowledged as it reflects a long-term, and still ongoing, research agenda on this topic. This contribution can be summarised in two main aspects. First, several of their studies highlighted the implicit contradiction, or dualism, of the CAP design and implementation (Esposti 2007, 2011b). Since its start in the sixties, the largest part of CAP expenditure (in fact, for many years, the only one) was concentrated on farmers income support either through market intervention (price support) or via direct payments. This major stream of the CAP (currently known as its "first pillar") constituted a main force in the direction of the over-simplification and loss of entrepreneurship of agriculture already underway in the NEC area. In practice, the CAP has represented one of the key facilitators of the abovementioned dualism of the NEC model: the level of unit support (that is, per hectare) was, and still is, higher in more urban areas and lower in the most remote and peripheral rural ones.

At the same time, however, in its initial intentions and even more clearly in the sequence of reforms started in the eighties, this EU policy was expected to provide a major impulse to the rural and agricultural renaissance contrasting that dualism, therefore to be the outpost of a necessary territorial and sectoral rebalancing. The birth of the Rural Development Policy in the late eighties (currently the "second pillar" of the CAP) was hailed as a critical breakthrough in this respect. The AAE deeply investigated this long internal struggle of the CAP between its conservation and its real reform and published many empirical studies clearly demonstrating how this dualism had a locally specific impact on agricultural activities, on its environmental implications and on the rural economy as a whole (Arzeni et al. 2001; Lobianco and Esposti 2010; Coderoni and Esposti 2018).

A second contribution of the AAE regarding the CAP was a deeper understanding of the forces actually leading to its design and implementation. In advance of at least one decade with respect to the broader debate on the EU weaknesses and uncertain future perspectives, they have shown how the EU policies actually are the outcome of a complex interaction between EU institutions, between Member States and between the former and the latter. This interaction eventually leads to a compromise that not only can explain the abovementioned dualism of the CAP, but also motivates the extreme complexity of its implementation within the local contexts. Several studies clarified how subsidiarity may eventually produce a highly complex governance of the EU policies whose possible outcome can also be their ultimate re-nationalization and, therefore, a substantial loss of significance of the EU dimension itself (Sotte 2005, 2007, 2010; Bonfiglio et al. 2016, 2017).

The bottom line of this research effort, in fact, goes beyond the relevance of the CAP itself. It has more to do with the need of appropriate theoretical backgrounds, approaches and methodologies in performing a rigorous policy evaluation. It is the local level, that is where beneficiaries and recipients actually operate, the context where policies should be actually evaluated but also where assessing the impact of a single policy is extremely complex. First of all, because its impact can never be isolated from the effects of the other concurrent (and sometimes contrasting) EU, national and regional policies. Secondly, because its impact depends on that sort of "collective intelligence" that is always strongly local specific and thus makes the impact of the policies substantially heterogeneous across space. The paradox, in this respect, is that several EU policies are aimed to build this local "collective intelligence" but their effectiveness in fact depends on its existence (Esposti et al. 2002; Arzeni et al. 2003; Camaioni et al. 2016).

The most recent contributions of the AAE concentrate on the theory and the practice of policy evaluation using the EU policies, and the CAP in particular, as a sort of privileged laboratory. Some relevant works, in particular, have applied to the so-called quasi-experimental ex post evaluation methods (Esposti 2017a, b). This research agenda is still trying to extend policy evaluation by also taking into account its implementation, the unintended and deadweight effects, its interaction with other policies, its influence on economic agents' behaviour. Taking all these aspects into account requires multiple and complementary methodological approaches on which the future research activity of the Ancona scholars is expected to provide further relevant contributions.

References

Anselmi, S. (1978). *Mezzadri e terre nelle Marche: studi e ricerche di storia dell'agricoltura fra Quattrocento e Novecento*. Bologna: Patron.

Anselmi, S. (2000). *Chi ha letame non avrà mai fame. Studi di storia dell'agricoltura, 1975–1999*. Proposte e ricerche. Quaderno monografico n. 26/2000.

Arzeni, A., Esposti, R., and Sotte, F. (2001). *Agricoltura e natura*. Associazione Alessandro Bartola. ISBN: 88-464-3169-3. Milano: Franco Angeli.

Arzeni, A., Esposti, R., & Sotte, F. (Eds.). (2003). *Politiche di sviluppo rurale tra programmazione e valutazione*. Milano: Franco Angeli.

Arzeni, A., Esposti, R., and Sotte, F. (2014). *Agricoltura e territorio: dove sono le imprese agricole?* QA—Rivista dell'Associazione Rossi-Doria, No. 1.

Baldoni, E., Coderoni, S., & Esposti, E. (2018). The complex farm-level relationship between environmental performance and productivity. The case of carbon footprint of Lombardy FADN farms. *Environmental Science & Policy, 89*, 73–82.

Bartola, A. (1979). Trasformazioni agrarie nelle Marche: un contributo interpretativo. *Diritto ed Economia, 2*.

Bartola, A. (1983a). L'agricoltura nello sviluppo economico delle Marche. In F. Amatori & G. Petrini (Eds.), *Problemi dell'economia e del lavoro nelle Marche*. Milano: Franco Angeli.

Bartola, A. (1983b). Agricoltura e sviluppo delle aree rurali: due comunità montane delle Marche. *Economia Marche, 2*.

Bartola, A., & Sgroi, A. (1984). Esperienze di programmazione a confronto. *La Questione Agraria, 16*.

Bartola, A., & Sotte, F. (1983). La programmazione in agricoltura: Riflessioni critiche suggerite dalle esperienze regionali. *La Questione Agraria, 10*.

Bartola, A., Sotte, F., & Sgroi, A. (1984). Il settore agricolo: problemi, prospettive e strumenti per la ristrutturazione produttiva. In F. Bronzini, & P. Jacobelli (Eds.), *Territorio montano e pianificazione operativa. Metodi, tecniche e politiche di intervento in una realtà comprensoriale dell'Appennino medio-Adriatico*. Milano: Franco Angeli.

Bonfiglio, A., Camaioni, B., Coderoni, S., Esposti, R., Pagliacci, F., & Sotte, F. (2016). Where will EU money go? Alternative policy scenarios for the re-distribution of CAP expenditure across the European space. *Empirica, 43*(4), 693–727.

Bonfiglio, A., Camaioni, B., Coderoni, S., Esposti, R., Pagliacci, F., & Sotte, F. (2017). Are rural regions prioritizing knowledge transfer and innovation? Evidence from rural development policy expenditure across the EU space. *Journal of Rural Studies, 53*, 78–87.

Buckwell, A., & Sotte, F. (Eds.). (1997). *Coltivare l'Europa*. Liocorno Editori: Per una nuova politica e rurale comune.

Calafati, A. G. (2008). Urban Sprawl Italian Style. *Scienze Regionali/Italian Journal of Regional Science, 7*(3), 99–108.

Camaioni, B., Esposti, R., Pagliacci, F., & Sotte, F. (2016). How does space affect the allocation of the EU rural development policy expenditure? A spatial econometric assessment. *European Review of Agricultural Economics, 43*(3), 433–473.

Coderoni, S., & Esposti, R. (2014). Is there a long-term relationship between agricultural GHG emissions and productivity growth? *A Dynamic Panel Data Approach, Environmental and Resource Economics, 58*(2), 273–302. https://doi.org/10.1007/s10640-013-9703-6.

Coderoni, S., and Esposti, R. (2018). CAP payments and agricultural GHG emissions in Italy. A farm-level assessment. *Science of the Total Environment, 627* (2018), 427–437. https://doi.org/10.1016/j.scitotenv.2018.01.197.

Esposti, R. (2000). Spillover tecnologici e progresso tecnico agricolo in Italia. *Rivista di Politica Economica, 90*(4), 27–78.

Esposti, R. (2002). Public agricultural R&D design and technological spill-ins. A dynamic model. *Research Policy, 31*(5), 693–717.

Esposti, R. (2004). Prospettive per lo sviluppo locale del territorio di collina e montagna delle Marche: alcune riflessioni. In Associazione "Alessandro Bartola" (Ed.), *Agrimarcheuropa. Una riflessione collettiva sulle prospettive a medio e lungo termine del sistema agricolo e alimentare delle Marche* (pp. 187–202). Milano: Franco Angeli.

Esposti, R. (2007). Regional growth and policies in the European Union: Does the Common Agricultural Policy have a counter-treatment effect? *American Journal of Agricultural Economics, 89*(1), 116–134.

Esposti, R. (2011a). Convergence and divergence in regional agricultural productivity growth. Evidence from Italian Regions, 1951–2002. *Agricultural Economics, 42*(2), 153–169.

Esposti, R. (2011b). Reforming the CAP: An agenda for regional growth? In S. Sorrentino, R. Henke & S. Severini (Ed.), *The Common Agricultural Policy after the Fischler Reform. National Implementations. Impact Assessment and the Agenda for Future Reforms* (pp. 29–52). Farnham: Ashgate.

Esposti, R. (2012a). Alcune considerazioni sulla retorica dello «sviluppo diffuso». In G. Canullo & P. Pettenati (Eds.), *Sviluppo economico e benessere* (pp. 285–305). Napoli: Saggi in ricordo di G. Fuà. Edizioni Scientifiche Italiane.

Esposti, R. (2012b). The driving forces of agricultural decline: A panel-data approach to the Italian Regional Growth. *Canadian Journal of Agricultural Economics, 60*(3), 377–405.

Esposti, R. (2012c). Knowledge, technology and innovations for a bio-based economy: Lessons from the past, challenges for the future. *Bio-based and Applied Economics, 1*(3), 231–268.

Esposti, R. (2014). On why and how agriculture declines. evidence from the Italian post-WWII period. *Structural Change and Economic Dynamics, 31,* 73–88.

Esposti, R. (2017a). The Empirics of decoupling: Alternative estimation approaches of the farm-level production response. *European Review of Agricultural Economics, 44*(3), 499–537.

Esposti, R. (2017b). The heterogeneous farm-level impact of the 2005 CAP-first pillar reform: A multivalued treatment effect estimation. *Agricultural Economics, 48*(3), 373–386.

Esposti, R., Arzeni, A., & Sotte, F. (Eds.). (2002). *European policy experiences with rural development*. Kiel: Wissenschaftsverlag Vauk.

Esposti, R., & Berloni, D. (2001). Scelte residenziali e delimitazione spaziale dei mercati locali del lavoro. Una applicazione alla Regione Marche. *Rivista Internazionale di Scienze Sociali, 3,* 291–312.

Esposti, R., & Bussoletti, S. (2008). Impact of objective 1 funds on regional growth convergence in the EU. A panel-data approach. *Regional Studies, 42*(2), 159–173.

Esposti, R., & Listorti, G. (2009). La competitività agroalimentare regionale. In A. Arzeni (Ed.), *Il sistema agricolo e alimentare nelle Marche. Rapporto* 2008 (pp. 339–368). Roma: INEA— Edizioni Scientifiche Italiane.

Esposti, R., & Materia, V. C. (2016). The determinants of the public cofinancing rate for applied R&D: an empirical assessment on agricultural projects in an Italian region. *R&D Management, 46*(S2), 521–536.

Esposti, R., & Pierani, P. (1995). Capacità Utilizzata e produttività dei fattori. Il caso di una impresa ex-mezzadrile. *La Questione Agraria, 60,* 71–99.

Esposti, R., & Pierani, P. (2003). Building the knowledge stock: lags, depreciation, and uncertainty in R&D investment and productivity growth. *Journal of Productivity Analysis, 19*(1), 33–58.

Esposti, R., & Sotte, F. (Eds.). (1999). *Sviluppo rurale e occupazione*. Milano: Franco Angeli.

Esposti, R., & Sotte, F. (2000). Società rurali, sistemi locali e mercati del lavoro. Una rilettura del caso marchigiano. *Economia e Società Regionale, 18*(1), 82–111.

Esposti, R., & Sotte, F. (Eds.). (2001a). *Le dinamiche del rurale. Letture del caso italiano*. Milano: Franco Angeli.

Esposti, R., & Sotte, F. (2001b). Institutional framework and decentralisation in rural development. In: C. Csaki, & Z. Lerman (Eds.), *The challenge of rural development in the EU accession process* (pp. 31–44). World Bank Technical Paper No. 504, Washington: World Bank.

Esposti, R., & Sotte, F. (Eds.). (2002a). *La dimensione rurale dello sviluppo locale. Esperienze e casi di studio*. Milano: Franco Angeli.

Esposti, R., & Sotte, F., (Eds.) (2002b). Institutional structure, industrialization and rural development. An evolutionary interpretation of the Italian experience. *Growth and Change 33*(1), 3–41.

Esposti, R., & Sotte, F. (2011). La ruralità nel futuro del modello marchigiano. In: Unioncamere Marche e Università Politecnica delle Marche (Eds.), *Le Marche oltre la crisi* (pp. 73–84). Milano: Franco Angeli,

Finocchio, R., & Sotte, F. (2006). *Guida alla diversificazione in agricoltura*. Ancona: Coldiretti Marche.

Fuà, G., & Zacchia, C. (Eds.). (1984). *Industrializzazione senza fratture*. Bologna: Il Mulino.

Lobianco, A., & Esposti, R. (2010). The Regional Multi-Agent Simulator (RegMAS): An open-source spatially explicit model to assess the impact of agricultural policies. *Computers and Electronics in Agriculture, 72*(1), 15–26.

OECD. (2006). *The new rural paradigm: Policies and governance*. Paris: OECD.

Orlando, G. (1965). *Programmazione regionale dell'agricoltura*. Argalia, Urbino: Studio pilota di piano zonale.

Sotte, F. (Ed.) (1987). *Agricoltura sviluppo ambiente. Una ricerca interdisciplinare sulle trasformazioni dell'agricoltura nelle Marche*. Istituto Gramsci Marche—Lega per l'Ambiente, Cooperativa Ecologia Editrice.

Sotte, F. (1996). Sviluppo economico, agricoltura e politica agraria. L'attualità del pensiero di Alessandro Bartola. *La Questione Agraria, 62,* 109–126.

Sotte, F. (1997). Per un nuovo patto sociale tra gli agricoltori e la società in Italia e in Europa. *La Questione Agraria, 65*.

Sotte, F. (2005). From CAP to CARPE: The state of the question. In K. M. Ortner (Ed.), *Assessing rural development policies of the Common Agricultural Policy*. Introductory Paper to the 87° EAAE Seminar, April 2004. Vienna, Kiel: Wissenschaftsverlag Vauk.

Sotte, F. (2006). Imprese e non-imprese nell'agricoltura italiana. *Politica Agricola Internazionale, 1,* 13–30.

Sotte, F. (2007). L'Health Check della PAC e il dopo-2013. Quali le poste in gioco? *Agriregionieuropa, 11*.

Sotte, F. (2008). L'evoluzione del rurale. Teoria e politica per lo sviluppo integrato del territorio. *Argomenti, 22,* 5–26.

Sotte, F. (2010). La politica di sviluppo rurale dell'UE. Riflessioni a margine del dibattito italiano. *QA—Rivista dell'Associazione Rossi-Doria, 1*.

Sotte, F. (2013). Scenari evolutivi del concetto di ruralità. *Proposte e Ricerche, 36*(71), 122–144.

Arbitration, Jurisdiction, Growth

Daniele Mantucci, Antonio Di Stasi, Laura Torsello, Alessandro Giuliani, Laura Trucchia, Mariacristina Zarro, Sara Zuccarino, Pietro Maria Putti, Alessandro Calamita and Christian Califano

Abstract An increasing interest for the operative and remedial aspects of justice administration has overwhelmed the labile distinction between substantive and procedural law. Furthermore, arbitration has always been spurring a fertile discussion among scholars from different fields of law, with different cultural backgrounds. For a long time, the Ancona Team of Lawyers considers arbitration law as an essential area in interest, also in light of the positive effects that the spread of arbitration would have on society, growth and competitiveness of Italian companies. Arbitration proceedings, in fact, represent a very important dispute resolution method, since it is

D. Mantucci (✉) · A. Di Stasi · L. Torsello · A. Giuliani · L. Trucchia · M. Zarro · S. Zuccarino · P. M. Putti · A. Calamita · C. Califano
Department of Management, Università Politecnica delle Marche, Piazzale Martelli 8, 60121, Ancona, Italy
e-mail: d.mantucci@univpm.it

A. Di Stasi
e-mail: a.distasi@univpm.it

L. Torsello
e-mail: l.torsello@univpm.it

A. Giuliani
e-mail: a.giuliani@univpm.it

L. Trucchia
e-mail: trucchia@univpm.it

M. Zarro
e-mail: m.zarro@univpm.it

S. Zuccarino
e-mail: s.zuccarino@univpm.it

P. M. Putti
e-mail: p.m.putti@univpm.it

A. Calamita
e-mail: a.calamita@univpm.it

C. Califano
e-mail: c.califano@univpm.it

© Springer Nature Switzerland AG 2019
S. Longhi et al. (eds.), *The First Outstanding 50 Years of "Università Politecnica delle Marche"*,
https://doi.org/10.1007/978-3-030-33879-4_7

83

different, but at the same time, complementary to State jurisdiction and other ADR tools.

1 An Introduction to the Study of Jurisprudence in the Polytechnic University of Marche

Daniele Mantucci

From Giorgio Fuà's point of view labour law and public law had a primary importance for economic studies, and this why these areas of law remained crucial for the lawyers of the Polytechnic University of Marche.[1]

The social and cultural evolution, however, suggested to widen the focus. In fact, the inner unity of the legal system and the progressive softening of the distinctions among different fields of law led to specifically address also civil law, commercial law, procedural law and, generally speaking, law and economics, as it is now a commonly accepted ground for studying economic transactions.[2]

According to the above, the Ph.D. program in *Law and Economics* was then established and it later became a characterizing element of the Ph.D. program in *Management and Law*. Ph.D. classes have been taught by Professors from the Polytechnic University of Marche, as well as by some very prestigious guests as Prof. Pietro Perlingeri and Prof. Antonio Baldassarre.

The complexity of juridical relationships has been analysed from the angle of constitutional values,[3] as to overcome the simplistic dichotomy between individual and collective interests, avoiding thus any formalistic application of the rule of law. Accordingly, a significant attention has been devoted to the asymmetries in parties' positions, such as when a consumer[4] or a weaker corporation[5] is involved in proceedings. This has led to deepen the knowledge of the market as it is the most important instrument supporting economic transactions but, at the same time, it is also aimed at the implementation of Constitutional values. Hence the law review *Persona, mercato, istituzioni* (The person, the market, institutions), presented in 2008 by Prof. Pietro Rescigno and honoured by the participation of prominent guests such as Alfonso Quaranta (the former President of the Italian Constitutional Court).[6]

---

[1]See Di Stasi A (2013, pp. 1–252), Di Stasi and Torsello (2015); Arethuse (2015); Torsello (2015). Regarding Public Law see Trucchia (2018).

[2]*Cfr.* Piperata and Trucchia (2016, pp. 97–117).

[3]See the fundamental reflections of Perlingieri (2006), *passim*, spec. p. 535 *et seq.*

[4]The general figure of the consumer is further articulated into different species, with different degrees of protection.

[5]Mantucci (2004), p. 52 *et seq.*

[6]Quaranta (2013).

The relevance accorded to economic transactions has produced interdisciplinary researches, such as those regarding intangible assets,[7] which were jointly conducted with scholars studying business organisation.

Furthermore, the importance of a concrete protection of these rights from possible infringements, has spurred the study of State jurisdiction as well as the evolving system of ADR methods.[8]

The traditional distinction between substantive law and procedural law is progressively fading away, and so does also the dichotomy between remedies based on party autonomy and those grounded on State jurisdiction.[9] And in such a context, arbitration law, as it is a fertile soil for discussion among civil law, procedural law, company law, international law, public law and tax law scholars,[10] has been thoroughly analysed.

Moreover, arbitration law induces to think to the effects that justice administration, generally speaking, has on society and, more in particular, on economic growth.

In fact, the inefficiency of State Court justice represents a heavy cost for corporations and constitutes a serious obstacle to the development, wellbeing and modernisation of the country.

Thanks to the studies of some of the most famous economist, such as Adam Smith and Max Weber, Oliver Williamson and Douglass North, it is now commonly accepted that the *wealth of nations* also depends on the judicial system and, most of all, on its efficiency in securing the performance of contractual undertakings. Consequently, alternative dispute resolution methods are gaining, or maybe re-gaining, a crucial role in the justice administration system.

Within the ADR galaxy, arbitration occupies a key position as it is a procedure with a jurisdictional nature, aimed to the settlement of concrete and practical disputes, fully compliant with due process requirements and held in front of private judges who will issue an award with the same effects of any other judicial decision rendered by a State Magistrate, which could be set aside in front of the competent Court of Appeals.

An old bias in favor of State monopoly of justice administration, characterized by a totalitarian vision of jurisdiction and public administration, has hindered the spread of arbitration,[11] even though the latter enjoys a prestigious and strong tradition in Italy.

The exhausting duration of State Court litigation and the reference to the principle of horizontal subsidiarity by the Italian Constitution have relaunched domestic arbitration.

The crisis that hit our economy has manifested itself in different shapes, around different parts of the world. Italian corporations are striving to find mew markets

---

[7]Zarro (2017); Zarro (2018) *et seq.*

[8]Mantucci (2008), p. 7.

[9]Regarding proedural agreements see Chizzini (2015).

[10]Califano (2004).

[11]For a general overview on arbitration in family law disputes see Putti (2014), *et seq.*; and as for disputes regarding electricity and gas retail industries see Zuccarino (2014) in *ivi.*

and insistently feel the need of a true globalisation. The spread of international transactions makes of arbitration an indispensable tool for the settlement of disputes which could not be otherwise efficiently handled by the domestic judicial system of a single State.

Arbitration, jurisdiction and growth: these are the characterizing elements of the research carried out by the Ancona Team of Lawyers, in collaboration of prestigious universities[12] and important law firms, with a specific focus on the increasing demand for justice coming as requested for a long time by the business and employment world.

2 Arbitration in Labour Law

Antonio Di Stasi

The relationship between arbitration law and labour law during the last century of the Italian history is a perfect angle to assess how the change in the social context reflect on the administration and enforcement of justice. Amid the high tides of democracy and the low tides of totalitarian regimes, some dramatic changes affected both the concepts of employment relationship and arbitration. From the liberal *Probiviral Committees* during the late nineteenth century, to the very restrictive approach during the fascist regime, the remedies available under employment law varied significantly, and the same happened to the relationship between employment law and arbitration law.

2.1 Arbitration at the Dawn of Labour Law. From the Pluralism of *Probiviral Committees* to the Establishment of a Special Judicial Branch During the Fascist Period

Antonio Di Stasi

Arbitration, as a mechanism for the resolution of disputes related to the execution of a contract, has its roots in contract law, as well as in labour law. In fact, when special legislation concerning labour law first started its growth, arbitration experienced a significant degree of development due to a strictly contractual approach to the employment relationship.

---

[12] See, in cooperation with the Dickson Poon School of Law of the King's College in London, Tavartkiladze (to be published). From the same Author, in cooperation with the Milan Chamber of Arbitration and Azzali (to be published). Gli arbitrati secondo regolamenti precostituiti. In Dittrich L (ed.), Trattato di procedura civile, UTET, Turin

As a matter of fact, the embryonic stage of development of labour law, characterised by an abstract and formal concept of a level playing field, incentivised the use of a private tool for the settlement of conflicting interests in matters which were mostly left to parties' contractual autonomy, with barely no peremptory or mandatory applicable rules.

To a deeper scrutiny, indeed, contract per se has a typically conciliatory nature, harmonizing parties' interests which are potentially always contraposed to each other.

In such context of extra-judicial resolution of employment controversies, arbitration and conciliation evolved together with the changing of the contractual relationships to which they referred.

During the dawn of labour legislation, therefore, arbitration law had to reconcile the two conflicting facets of being an instrument with a strong contractual nature and being, at the same time, forced to adapt itself to a changing reality that did not necessarily conform itself to the classic rules of the civil code.

However, it must be said that the so called «private justice», a justice which is alternative to the one delivered by the State, evidently suffered from the circumstance that it had its origin and source in a private agreement and was administered by private individuals with no judicial power, in the strict sense of the term. The demand for a more far-reaching party autonomy, typical of a liberal prospective, was thus limited by the State reserving for itself, in a significant though not exclusive basis, the power of *ius dicere*.

In light of the above, arbitration may be said to have both a contractual origin—resembling thus some of the features of an arbitrage, with a third person delegated by the parties to regulate a certain relationship—as well as a slightly jurisdictional nature, which makes it an alternative instrument for the enforcement of justice, much faster and flexible than litigation before State Courts.

In such a context, collective bargaining of compensation and rights has considerably contributed to the development and to a better understanding of arbitration.

Clear examples of this contribution are given by arbitration commissions such as the one constituted in Milan, in 1880s, within the printing industry, or the arbitral jury dated 1883, instituted within the silky industry in Como, with the attached arbitral panel operating therein.

The old *agreements on compensation* [*concordati di tariffa*, in Italian], ancestors of the modern collective agreements, contemplated arbitral tribunals dealing with controversies internal to the guilds of bakers in Monza, printers and gravers in Milan.

Such kind of arbitrations were also aimed at clarifying and implementing the set of rules agreed upon by the employer and the employees, and the more bargaining power trade unions had, the more these arbitrations were developed. Indeed, arbitral tribunals were envisaged in the 1906 Itala–Fiom collective contract, as well as in the collective agreement between Borsalino industries and the Italian federation of hat makers.

The competencies and skills developed therethrough arbitration encouraged the further establishment of special *probiviral committees*, which were introduced by Law n. 295 on 15 January 1893, and the following Regulation n. 179, enacted on 26 April 1894. These *probiviral committees* produced a remarkable jurisprudence and represented an example of a special pool of magistrates, paving the way for the establishment of a special judicial branch dealing with labour law, on an exclusive basis.

Probiviral committees, in facts, envisaged the participation of specialists in labour controversies, with a different cultural and technical background from that of ordinary magistrates.

However, the worst limit of *probiviral committees*, from a systematic point of view, was the exclusion of collective controversies from the scope of its jurisdiction, as State Courts had instead an exclusive jurisdiction on these disputes.

During the first post-war years, the rise of the fascist party determined a significant turnover in the legislative policy, with a remarkably high degree of aversion against working class, to the extent that since the Palazzo Vidoni agreement, only fascist trade unions could be recognised, while factory's internal commissions were abolished and, later on, with Law n. 563, in 1926, striking and shutdown were declared criminal offences. So that, even though *probiviral committees* have been progressively abandoned in favour of a special judicial branch, with an exclusive jurisdiction on labour law controversies, nonetheless, such *probiviral committees* have been a positive development, in line with the trend of the time.

On the other hand, however, justice administration clearly limited the access to State Courts for collective claims by requiring that fascist trade unions had to join the process, which was very unlikely to happen since the only existing trade unions were those officially recognized by governmental authorities, with only one association for each category of employer and employees in every different field of industry.

Even though Law n. 563/1926 was still in force—and so were also the *probiviral committees* and the district arbitration commissions for private employment, whose decisions were to be appealed in front of the competent Court of Appeals—the enactment of Royal Decree n. 471/1928 and the following Royal Decree n. 1037/1934 had a double effect. On one side, they strengthened amicable dispute settlement mechanisms as conciliation became a mandatory procedural requirement which had to be met before the claim could be brought before State Courts. On the other side, however, arbitration clauses were banned from collective agreements under sanction of nullity and the 1940 Italian code of civil procedure finally prohibited arbitration clauses in individual employment contracts, agreements on compensations, as well as in any other equivalent or similar source.

Parallelly, jurisdiction on controversies involving labour law was vested on district judges known as *Praetors*, or local Tribunals, depending on the value of the dispute, so that the disputes could be decided after short and concise proceedings. Nonetheless, the requirements for the valid composition of the panel created some relevant difficulties, as the final decision had to rendered immediately after the discussion on the merits of the case by three judges, two of whom were the Chairman presiding the first hearing and the judge delegated to the taking of evidences.

With the enforcement of law under the tyrannical monopoly of a totalitarian State, since there were no other means for administering justice, the centralisation plot of the fascist legislator was fully displayed. As a consequence, even though procedural reforms enacted in 20s and 30s introduced some positive changes in favour of employees, the overall protection was significantly worse off due to the gap between the way justice was administered and the reality of labour disputes. On these bases, the 1940 Italian code of civil procedure set forth an entire book regarding individual employment and social security claims, as well as collective controversies, thus unifying in itself all the proceedings regarding employment disputes. By contrast, this meant the end of any alternative dispute settlement mechanism.

2.2 Arbitration in Labour Law in the Republic Constitution: The Reasons Behind the Success of Art. 7 of the Employees' Charter

Laura Torsello

After the dismantlement of the fascist legislation and the entry into force of the Republican Constitution, the renaissance of labour law developed for the most part around Law n. 604/1966 and Law n. 300/1970 [also known as the *Employer's Charter*]. These statutes introduced an important set of innovative substantive rules which were, at a later stage, complemented by the procedural rules introduced with Law n. 533/1973, concerning individual employment and social security claims. Indeed, Art. 7 of Law n. 604/1996 provided for arbitration proceedings against unfair or wrongful dismissals, ending with a non-enforceable award (*arbitrato irrituale*); while Art. 7 of Law n. 300/1970 limited employers' rights in the exercise of disciplinary power and set forth formal requirements to be met in order to do so. At the same time, Art. 28 of Law n. 300/1970 introduced very fast and effective proceedings for claims concerning alleged violations of trade unions' rights or infringements of the right to strike.

Arbitration under Art. 7 of Law n. 300/1970 is particularly interesting for the present purposes as it is aimed at the settlement of disciplinary claims in respect of any disciplinary decisions heavier than written warnings, including among others, as clarified by the constitutional Court: suspension from work without pay, dismissal for justified objective reasons, as well as dismissal for good cause.

In its relevant part, Art. 7 of the Employee's Charter allows the employee, together with other possible solutions, to file a request of arbitration with the local labour commission within twenty days from the application of the disciplinary sanction, asking for the constitution of an arbitral panel composed of one arbitrator for each of the parties and a third arbitrator, jointly appointed by the parties, or failing such agreement, by the Director of the local labour commission.

So that, even though the time limit for the commencement of arbitration may lead employees to prefer ordinary litigation instead of arbitration, for the first could be initiated within a longer period, the fact that the sanction is immediately suspended, since the commencement of arbitration and until the arbitral panel reaches a final decision, encourages to use arbitration.

Moreover, there are two other important rules in favor of the employee that need to be considered. In fact, in the case the employer fails to appoint its arbitrator within ten days from the receipt of the arbitration notice from the labour commission, the sanction immediately ceases its effects. Furthermore, should the employer bring a lawsuit in front of a State Court, the sanction would nonetheless be suspended until a final decision is reached.

Law n. 533/1973 developed some of the ideas envisaged *in nuce* by the fascist legislation but, unlike the latter, did not impose any governmental policy inspired by the battle between capital and labour. Instead, employees were granted with the right to fast and efficient proceedings, for an accurate and reliable ascertainment of the facts giving rise to the dispute. So that, alongside with the ordinary (but sometimes too formal) proceedings, an entirely new kind of procedure had been introduced, with rigorous preclusions applying to the definition of the subject matter to be decided and to the taking of evidences.

On the other side, the new procedure outlined by Law n. 533/1970 is a faster means for the decision of the dispute, grating due process since the fixing of the first hearing, throughout the entire proceedings, until a final decision is reached after the parties had a reasonable opportunity to present their cases.

The goal pursued with the introduction of these procedural rules was that of reducing the time needed to reach a final decision, thanks to a series of rigorous preclusions and an increase of the judge's inquisitorial powers, so as to enhance the principles of orality, concentration and immediacy, according to the theories of Giuseppe Chiovenda that inspired the Italian code of civil procedure. All the more so, the entire process, for instance, could finish with a decision being adopted soon after the hearing for the taking of evidences and the discussion on the merits of the case.

2.3   The New Paradigms of Labour Law and the (Attempted) Overcoming of State Jurisdiction in Favour of Arbitration

Alessandro Giuliani

The increase in the rights and protection accorded to employees paired with the progressive consolidation of the role of State Courts within the realm of the remedies provided by labour law. Nevertheless, the last decades of the 20th century, characterised by the collapse of the socialist system in Eastern Europe and the shadows of a revolution fading away, marked a significant decrease of the welfare and social security levels, giving an ideological ground to attacks against the basic structures of labour law.

From the prospective of the levels of protection accorded to employees, the new paradigms are given by key concepts such as: flexibility, saving costs and less effective guaranties. Parallelly, from a procedural point of view, the focus is back again on arbitration for the settlement of employment actions, in lieu of State Court litigation.

Art. 31 of Law n. 183/2010 rewrote Art. 410 of the Italian code of civil procedure with the aim of enhancing private justice and encouraging the use of alternative dispute settlement mechanisms, instead of State Court litigation. There are now 4 different kinds of arbitration in labour law, each governed by specific provisions with an undesirable proliferation of employment arbitrations, contradicting the aim of simplifying proceedings and, rather on the opposite, not even trying to coordinate such proceedings.

On the other hand, however, it is worth observing that all these arbitration proceedings lead to awards having contractual effects as per Art. 1372 of the Italian civil code and, at the same time, are characterized by a rather jurisdictional nature (and are often articulated in two phases, combining conciliation to arbitration). These proceedings are to some extent similar to the *lodo irrituale* provided for by art. 808 *ter* of the Italian code of civil procedure, but in respect of some peculiar aspects they depart from the ordinary rules.

Given that employment arbitration under Art. 412 of the Italian code of civil procedure may be initiated at any time by the parties to conciliation proceedings in front of the same conciliation commilssion dealing with the case, the same should apply also to the procedures under the auspices of the arbitration chambers attached to certifying legal bodies, as provided by the above-mentioned Art. 31, paragraph 12.

Art. 412 *ter* of the Italian code of civil procedure contemplates another kind of employment arbitration which follows the rules established in the collective agreements signed by the most representative associations representing employees. Remarkably, this provision just refers to collective agreements and does not put any limit to the arbitrability of employment disputes, nor affects the nature or the effects of the award, as well as the availability of remedies to set it aside.

Art. 412 *quarter* of the Italian code of civil procedure establishes a third kind of employment arbitration assuming that a dispute between the parties is already pending and dictates for rigorous time limits upon the parties, in a way that makes it somehow resemble the first kind of proceedings. Impartiality and independence of the arbitral panel are nonetheless guaranteed, for the parties may appoint arbitrators of their own choice and the third arbitrator, acting as chairman of the arbitral tribunal, is designated among law professor with a right of audience before the Court of Cassation by the party appointed arbitrators.

The fourth kind of employment arbitration is outlined by the abovementioned Art. 31, paragraph 10, and it allows the employer and the employee to agree on an arbitration clause pursuant to art. 808 of the Italian code of civil procedure. It further provides that the arbitral panel may decide the dispute *ex aequo et bono* and, as to the procedural aspects, refers to the provisions discussed above, regarding the first and the third kind of employment arbitrations.

So that, if the real aim pursued through these reforms was that of enhancing arbitration instead of state court litigation, it is quite easy to notice that such a variety

of proceedings could result in a useless complication of the procedural framework—
which is quite disappointing since arbitration for its own nature should be simple, fast
and effective. A good example in this sense is provided by the arbitration proceedings
under Art. 7 of the Employer's Charter, which is slightly less effective than State Court
litigation, but accords to employees some very significant protections.

3   Arbitration and Public Administration

Laura Trucchia

To analyse arbitration proceedings with public bodies means to deal with fundamental
issue in Public Law and particularly for Administrative Law.

Admitting that also controversies with public bodies are arbitrable means to put
in doubt the way how public rules are enforced, as well as the position of private
citizens in respect of Public Administration. In fact, arbitrability means that the
peremptory and coercive character of public enforcement is replaced by a more
consensual method. Moreover, the same relationship between a private citizen and
Public Administration is affected, since arbitration implies a more level playing
ground than that usually accorded to the *weakened* interests that a private could have
against Public Administration.

Furthermore, the same idea of State sovereignty is concerned when arbitration
is under the scrutiny from a public law prospective in so far jurisdiction is no more
conceived a State prerogative. Instead it is administered by privates according to the
rules and the principles enshrined in Art. 24 and Art. 102, as well as in Art. 118 of
the Italian Constitution regarding the principle of subsidiarity.

This is why these arguments stimulate the interest of both public and private law
scholars, who find many aspects to deal with and discuss, such as the many reforms
in the organization and operation of Public Administration, as well as privatisation.
The latter, in fact, has changed the relationship between privates and Public Administration and made it necessary to constantly monitor the ambiguous boundaries of
what is public and what is not and to make it sure that the choice between public law
and private law remedies does not jeopardize the fulfilment of the aims pursued with
the new regulation.

The choice to devolve a dispute to the jurisdiction of an arbitral tribunal implies a
double folded evaluation. The first aspect to be considered is the way public bodies
perform their duties, which per se is something totally permissible since Public
Administration is meant to serve public interests and is thus limited and capable
of being controlled. The second element under scrutiny is how public duties are
now being fulfilled through an administrative procedure, which may end with an
administrative decision or even with an amicable agreement.

A further element which deserves attention is the traditional distinction between
subjective rights and legitimate interests, even though they are sometimes interpreted
as to resemble each other, and other times as instead being radically different. Despite

such divergencies, it is generally agreed that, in both cases, it should be kept in mind that the relationship between privates and Public Administration is concerned and that both subjective rights and legitimate interests are concepts which constantly evolve in time.

The issue concerning the arbitrability of a dispute on a legitimate interest against Public Administration also raises doubts as to the arbitrability of non-disposable subjective rights. Moreover, on a different level, a survey on the implications that arbitration law bear on public law affects the whole system of the remedies available against Public Administration. And this adds on top of the reform in the way State jurisdiction on administrative disputes is split between civil and administrative Courts. In fact, the traditional criterion for the determination of the competent Court based on the distinction between subjective rights and legitimate interests has been abandoned, in order to increase the scope of the exclusive jurisdiction of Administrative Courts. In so doing, parallelly, due process guarantees in favour of the private citizen have been increased and the proceedings have begun to deal with the entire relationship between the private citizen and Public Administration, instead of limiting the focus on a single administrative act.

Despite such news previsions, however, the administrative judge doesn't look to be fully able to protect privates and the very same newly introduced administrative process is not always an effective tool for the settlement of disputes.

Art. 12 of the Italian Code of the Administrative Process, in this regard, highlights that arbitration is an alternative mean for the settlement of disputes, meant to overcome the inefficiencies of State Court litigation and, furthermore, clarifies the scope of arbitrable disputes by pointing out that only disputes concerning subjective rights can be decided through arbitration. Even tough, due to the unclear boundaries of what may be defined as public and what may not, as well as to the increase in the role of party autonomy in administrative law, the issue of arbitrability needs to be addressed trying to enhance more effective tools and overcoming current pitfalls.

The overall context is therefore undergoing a deep renovation, following the wave of the European rules aimed to increase efficiency of proceedings but, on the other side, there is still a retrograde and regressive trend.

Conclusively, it is beyond any doubt appropriate to study arbitration law from a public law prospective not only on a domestic, but also on the international level, with a particular focus on the admissibility of arbitration in specific sectors, such as: public procurement and mediation procedures in front of independent administrative bodies—especially since doubts and objections have been raised as to arbitration proceedings on these matters.

4   The Present and the Future of Arbitration with Independent Administrative Authorities

Mariacristina Zarro and Sara Zuccarino

Independent administrative authorities first appeared in the 70s and in the next two decades have progressively spread through the entire legal system so that, thanks to the impulse from the European Union, they have even been endowed with a «quasi-judicial» power in order to carry out their functions, creating thus a hybrid entity, halfway from adjudication to enforcement of law.

Even though these independent authorities are commonly defined as «arbitrators of the market», their nature still needs to be attentively assessed. In fact, it is essential to clarify the relationship between proceedings before independent administrative authorities and arbitration, as well as to examine remedies available against the measures adopted by the same authorities.

The facts that each independent administrative authority is unique in its kind, and the remedies available thereunder are akin to no other, suggest that a dogmatic approach is not advisable. Indeed, administration of justice in these contexts is very fluid, since parties may avail themselves of mediation proceedings, arbitration, as well as other forms of alternative dispute resolution mechanisms. So that, even though a *sanction* imposed by an independent administrative body may very well be considered as a legal action, very different considerations apply in respect of those measures adopted in order to settle disputes between private parties. Such measures, indeed, depending on the specific circumstances of the case, have a jurisdictional or a quasi-jurisdictional nature and, this is the reason why, they shed a new light on the way independent administrative authorities carry out their functions.

Mediation proceedings are generally preferred to adjudicatory methods, however, when it comes to the requirements needed to access such mediation proceedings, there is some resemblance with arbitration proceedings.

Having said this, in first place, it is crucial to clarify that, despite the name, neither proceedings before the *Arbitro Bancario Finanziario* [Arbitrator for Banking and Finance disputes, which is a body attached to the Italian Central Bank], nor the *Arbitro per le controversie finanziarie* [Arbitrator for financial disputes, which is a body attached to the Consob, i.e. the Italian securities and exchange commission], are not really akin to arbitration. As a matter of fact, these proceedings, apart from the name, have not the features required to be considered as an arbitration, since decisions therefrom are not binding, financial intermediaries are obliged to agree to the commencement of these proceedings and, furthermore, the adjudicating panel is permanent, rather than being nominated for the single controversy.

The more interesting proceedings are those available before the *Autorità garante per le comunicazioni* (*breviter* AGCOM, i.e. the Italian Regulatory Authority for Communications) and the *Autorità di regolazione per Energia Reti e Ambiente* (breviter ARERA, i.e. the Italian Regulatory Authority for Energy, Networks and the Environment).

Both the administrative authorities have set forth proceedings for the settlement of disputes between corporations (B2B disputes), as well as those between a corporation and its customers (B2C disputes). Both proceedings consist of two phases: the first one is mandatory and has a conciliatory purpose; the second one is conditional and depends on the decision of the customer only. Such proceedings are characterized by the fact that the independent administrative authority has the power to adjudicate the dispute, and its *decisum* is fully binding upon the parties.

Making a specific reference to ARERA, the peculiarity of those arbitration proceedings deserves to be highlighted.

Arbitration for energy disputes was first conceived by Law 481/1995 but, after as little as 24 years, the necessary implementing regulation is still lacking. It was only at a later stage, with Resolution n. 127/2003 and Resolution 42/2005, that the old previsions have been abrogated and a new set of detailed and specific rules became available. Furthermore, these rules limited the scope of application of proceedings to only those disputes regarding electricity transmission services and gas transportation through pipelines.

Such kind of proceedings consist of an administered arbitration, as per Art. 832 of the Italian code of civil procedure, where the Regulatory body is also at the same time the administering institution, based on the reference contained in the arbitration agreement. As per Resolution 139/2006, the arbitral panel is composed of three arbitrators, each party may appoint one arbitrator, while the third arbitrator, acting as chairman of the arbitral tribunal, shall be appointed by the same Regulatory authority, after having considered the proposal by the General Director.

Despite the huge advantages offered by this kind of arbitration, such as rapidity, a specialist knowledge of the field, the stability of arbitral awards if challenged, based on the caseload of the Courts of Appeals, it seems that only one arbitral award has been rendered.

The reason why these proceedings are far from being widespread is to be found in the role of the regulating authority, with not simply administrative functions, but generally pursuing regulatory policies while dealing with the single specific case. This is the so-called «regulation by litigation», which, however, also jeopardises the outcome of the specific dispute involved in the proceedings.

Such an invasive role of the Regulatory Authority is a strong deterrent, especially for the owner of the network who fears that data and information disclosed during proceedings could be used against itself by the same Authority to charge possible sanctions. In other words, the need for a mutual agreement on arbitration, in this specific context, is a big obstacle to its development.

Other adjudicatory methods too, though different from arbitration, show all their limits and shed doubts as to the neutrality and impartiality of the administrative authority exercising its control on the final decision. From one hand, in fact, adjudication of the case is mixed with the regulatory functions of the administrative body; on the other hand, the setting aside of such decisions belongs to the jurisdiction of Administrative Courts, with all the consequential limits as to the scope of the judicial control on administrative activities.

Given the functions of these administrative independent authorities, which basi-cally have a jurisdictional functional too, the interpreter is left with the ques-tion whether proceedings under the auspices of these institutions comply with the constitutional and international requirement of due process.

Conclusively, the importance of the industries where independent administrative authorities are operating, their specialist knowledge of the market, the rapidity and cost-efficiency of proceedings, suggest to continue spurring ADR methods. Nonethe-less, criticisms as to the existing remedies should guide the legislator on how to improve the system.

5   Arbitration Law in Specific or Peculiar Contexts. Family Law

Pietro Maria

It is a truly consolidated idea that State Court litigation is not the best way to handle family disputes, which for their own nature do not fit in such proceedings. Court litigation, in fact, ends with the winning and losing parties, which is something that is not very useful for family members to settle divergencies in a positive way, in a way that could meaningfully contribute to a better life. Quite on the opposite, State Court litigation very often dissolves family ties, most sadly those between children and parents, and increases conflicts, rather than making family members caring for each other.

A professional family mediator would certainly do a better job than a State Court Judge, for the first has the technical instruments needed to help parties to talk to each other in a meaningful way. Furthermore, the mediator will not deliver any formal decision on who is right and who is not. Nonetheless, disputes involving family law often offer no space in civil law jurisdictions for party autonomy, as the subjective rights involved therein are commonly of a non-disposable nature. So that, it is of little surprise that family mediation has achieved its best results in common law systems, where marriage is conceived simply a contract. However, the new trend towards a more active role of the parties in solving family disputes has opened some new interesting scenarios.

The first-time family mediation had been envisaged in the European Context was with the enactment of the UN Convention on the rights of the child, which has been adopted by the Council of Europe in Strasbourg on 25 January 1996, and the following Regulation n. 2201/2003, adopted on 27 November 2003. Mediation has been conceived to help the competent national authorities to collaborate with each other regarding parental responsibilities while, on the national level, it was meant to enhance a cooperative and talkative approach towards family disputes, instead of juxtaposing conflicting interests, as Court litigation does. However, the path still seems very long and full of obstacles.

In fact, a new law proposal has been scheduled with the 2019 legislative agenda of the Parliament, aimed to discipline issues related to shared-custody and make of family mediation a mandatory requirement in order to bring a lawsuit before a State Court, if the dispute concerns, even indirectly, children's rights. However, despite the importance of the goals pursued, such as that of supporting families and children in need, currently, there is no legal provision really clarifying the way family mediation works.

So that the discipline of family mediation is limited to the provisions set forth by art. 342-*ter* and art. 337-*octies* of the Italian civil code. The first one, provides that the Judge may issue an order of protection from family abuses and mandate social services or family mediation centers *in loco* to intervene, whenever appropriate. Art. 337-*octies*, paragraph 2, of the Italian civil code, prescribes that «whenever appropriate, the Judge, after having heard the parties and having obtained their consent, may postpone the adoption of the measures described in Art. 337-*ter*, in order to allow spouses to try to find, with the help of an expert, a settlement agreement in the best moral and material interests of the children». And it is to be noticed that the way family mediation relates to State Court litigation shows how the first proceedings differ from other amicable dispute settlement instruments.

In this sense, Art 337-*octies*, paragraph 2, of the Italian civil code, describes family mediation as an eventual phase of separation legal proceedings. According to Art. 337-*octies* the decision to mandate parties to family mediation rests upon the discretionary evaluation of the Judge who, in so doing, would suspend legal proceedings in front of him and would defer the adoption of any further decision. Once such a decision is taken, the Judge will invite the parties to personally attend mediation hearings.

In fact, only after personally meeting with the parties the judge will be able to confirm or not his original intendment to mandate for family mediation with the assistance of a specialist. Moreover, during that hearing, parties will be asked to confirm whether they accept the invitation to attend family mediation proceedings. This is one the of the most peculiar aspects of the family mediation which, unlike other forms of mediation, is entirely based on parties' consent. Accordingly, it may be said that these proceedings are alternative to State Court trials, but they nonetheless prelude to a final decision by the Judge complying with party autonomy, with no particular formality. In this regard, it is understood that the Judge shall be precluded from scrutinizing any other aspect of the agreement but the moral and material best interest of the children, as well as other issues relevant for public policy purposes. Briefly, it is quite clear that family mediation cannot replace State jurisdiction, as the latter is the only venue for disputes regarding non-disposable rights. The rationale behind the system is that State is responsible for the protection of those rights that cannot be adequately protected in a free market system because of the very different bargaining power of the parties. So that, it is quite relevant to verify whether there is any room for party autonomy in family law, as already suggested by some authoritative civil law scholars.

Even though this is not the most appropriate place to discuss such complex arguments, however, it would be fair enough to hold that a paternalistic approach, detrimental to party autonomy, is of no advantage in protecting the weakest. The fact that jurisdiction on family law disputes is vested upon State Courts is the logical consequence of the Constitutional understanding of the household as an autonomous subject of law, distinct from the family members composing it, where party autonomy plays a less relevant role. Furthermore, parties' lack of a genuine contractual autonomy in family law controversies is also justified by public policy concerns, which evidently exceed parties' control, and, from a the more general point of view, by the inappropriateness of a contractual approach to Italian family law.

However, times have radically changed. The introduction of divorce has abolished the principle of indissolubility of marriage and the same idea that no one can bend to his convenience the legal effects stemming out of marriage. So that it is now time to redraw the boundaries of non-disposability in family law controversies. From this point of view, the new legal context and the jurisprudence on family law suggest the adoption of a different approach from the past, when only the State could discipline family matters, and let parties' agreement have a more important role. In other words, spouses should be free to agree on their respective rights and duties and even derogate the ordinary legal provisions, to the extent that they do not negatively affect the best interest of the household and of the children.

6 Sport and ADR

Alessandro Calamita

The Italian National Olympic Committee, on 15 July 2014, has approved a revolutionary reform of sport justice aimed to rationalise dispute resolution proceedings and correct shortcomings, as suggested by both practitioners and scholar, as to make justice administration in sport law more reliable.

Before the 2014 reform, jurisdiction was split in two branches. The High Court of Justice had jurisdiction concerning disputes over *non-disposable* rights; while, the National Sport Arbitral Tribunal could validly decide disputes involving *disposable* rights.

The new reform radically changed the system and introduced a General Attorney with the task of coordinating district attorneys' offices, eliminated the National Sport Arbitral Tribunal, replaced the High Court of Justice with the Guarantee Committee for Sports, leaving to each National Federation the decision on whether to introduce or not internal arbitration proceedings for «merely pecuniary disputes» occurring within the same federation (Art. 3, paragraph 3, of Sport Justice Code).

In the international context, nothing seems to have changed since the Tribunal Arbitral du Sport [hereinafter, TAS], based in Losanne, is still in force. TAS applies the Swiss Law according to Art. 177 of the Federal Act on Private International Law and has jurisdiction on «toute cause de nature patrimonial».

Accordingly, both on the national and the international level, availability of arbitration for sport-related disputes depends on the patrimonial nature of the claim, rather than on the non-disposability of the subjective right involved in the dispute. The patrimonial nature of the dispute is thus meant to replace the classic concept of non-disposability. However, it is far from clear whether this new criterion will promote legal certainty in assessing whether a certain controversy is arbitrable or not.

Both the TAS and the Swiss Federal Tribunal (the appellate court for TAS arbitral awards) confirmed that the criterion based on the patrimonial nature of the dispute has a wider scope than the one base on the disposability of the subjective right, enlarging thus the boundaries of arbitrability. In fact, a dispute may be said to have a patrimonial nature if the claims brought by the parties are meant to protect their economic interests.

Art. 4, paragraph 3, of the above-mentioned 2014 Sport Justice Code, instead speaks of merely patrimonial disputes. According to the new rules, in short words, only disputes internal to each federation could be considered arbitrable and the same disputes could also be brought before the Fourth Division of the Guarantee Committee which, in fact has jurisdiction on «merely patrimonial dispute», as per Art. 56, paragraph 2, letter d), of the Sport Justice Code. On the opposite, disputes dealt with by the first three Divisions of the Guarantee Committee cannot anymore be considered to be arbitrable, as it was before the 2014 reform when the National Sport Arbitral Tribunal could validly administer the same proceedings. This is the case, for instance, of dispute concerning disciplinary sanctions which now belong to the jurisdiction of the Second Division of the Guarantee Committee.

It is still too early to say how these new rules will concretely be implemented by national federations. However, the fact that concurrency of remedies only applies in respect of disputes internal to a single sport federation, i.e. to less important disputes, should not, unlike in the past, impact on proceedings in any significantly negative way. Furthermore, availability of arbitration for merely patrimonial disputes depends on the decision of the sport federation to insert an arbitration agreement in its statute. On the opposite, the now abolished National Sport Arbitral Tribunal had a much wider jurisdiction, since the arbitration agreement was set by the statute of the Italian National Olympic Committee.

There is another element conditioning the availability of arbitration proceedings for such kind of merely patrimonial disputes, internal to each single federation. In fact, according to commissioners specially appointed by the Italian National Olympic Committee to supervise the implementation of the new internal statutes of national sport federations, the arbitration agreement pursuant to Art. 4, paragraph 3, of Sport Justice Code is not binding upon the party who joined the federation. In other words the explicit or implicit acceptance of the statute in order to join the sport federation does not deprive that party of the possibility to access State Courts, as the arbitration agreement eventually provided for by that very same statute is to be considered as a mere proposal.

Scholars have criticized this approach highlighting that it goes against the basic rule provided by Art. 1367 of the Italian civil code, concerning interpretation of the contract. Art. 1367, in fact, provides that, where there is ambiguity and more than one possible interpretation, contract should be interpreted as to produce effects, rather than not producing any. While, this would be exactly the case if arbitration agreements pursuant to Art. 4, paragraph 3, of Sport Justice Code were interpreted as to be non-binding arbitration agreements. No one, in fact, has every cast any doubt on the fact the Sport Federations could, at any time, propose arbitration to counterparties, instead of State Court litigation, if the subject matter of the dispute was arbitrable—which is just the case of the «merely patrimonial dispute» referred to by Art. 4, paragraph 3, of Sport Justice Code.

A similar restrictive interpretation reminds to some extent the jurisprudence limiting availability of the *irrituale* arbitration (i.e. of arbitration proceedings leading to a non-enforceable award, a kind of arbitration peculiar to only Italian law) but is, under no circumstances, desirable for arbitral awards which, as per Art. 824 bis of the Italian code of civil procedure, enjoy the same effects of the decisions rendered by State Courts. All the more reason so why the scope of arbitration proceedings is now further increased by Art. 808 *bis* of the Italian code of civil procedure, stating that also non-contractual disputes may be decided by arbitral panels.

The 2014 reform has definitively downsized the role of mediation proceedings. Under the previous system, mediation proceedings had to have been tried before the claim could be brought in front of the National Sport Arbitral Tribunal, before it was abolished. Under the rules currently in force, mediation is merely eventual, poles apart from the general trend in other fields of law where mediation is generally conceived as a mandatory preliminary step, especially after the enactment of the Law n. 98/2013.

Despite such step back, several federations provide for mediation as a preliminary requirement before the commencement of arbitration and, moreover, the same arbitrators may exercise their powers as to invite parties to ascertain whether the dispute could be amicably settled.

While, in so far international sport law is concerned, TAS Arbitration Rules contemplate special mediation proceedings based on a prior agreement to mediate, in which each party undertakes to attempt in good faith to negotiate with the other party, in order to settle a sport dispute, with the assistance of a professional mediator.

Conclusively, sport law remains a field where patrimonial disputes are far from being unlikely. In order to deal properly with the peculiar aspects of such kind of disputes, alternative and flexible settlement methods are of the utter importance. The School of Economics of the Polytechnic University of Marche will continue being keen on studying these subjects.

7 Arbitration, Tax Law and Economic Growth

Christian Califano

Arbitration and tax law bear a significant importance for a survey regarding efficiency in the administration of justice.

There are several profiles to be considered. The first one concerns registration tax (due for arbitration agreements and arbitral awards) and the stamp duty (due for parties' briefs), for these taxes are requested in view of the nature of the proceedings to which they apply. Such a legal framework, however, has been «recently and repeatedly changed, as taxation now incentivizes arbitration in a trend towards the so called *dejurisdictionalization*, i.e. the progressive abandonment of State Court jurisdiction.

The system of fiscal incentives contemplates, among others, a special tax credit that parties can set off from the fees due to arbitrators, if proceedings end with a final award, as well as tax credits in case of negotiations with the assistance of parties' counsels, and other incentives for settlement agreements in civil and commercial disputes.

However, even though arbitration is enhanced through tax offsets, it may not be considered as a tool for dealing with tax obligations.

In such a context of *dejurisdictionalization*, striking an equitable balance between taxpayers' interests and those of the Fiscal Administration is of the utter importance. Accordingly, certainty and rapidity in reaching a final decision have to face the difficulties lying with the non-disposable nature of tax obligations.

From a tax law and economic perspective, the importance of alternative dispute settlement methods for controversies between taxpayers and fiscal administration is clearly witnessed by the 2017 *OECD Explanatory Statement to the Multilateral Convention to Implement Tax Treaty Related Measures to Prevent Base Erosion and Profit Shifting*, and by *Mutual Agreement Procedures*—i.e. amicable settlement procedures available to taxpayer in order to avoid double taxation, should two or more States impose their taxes upon the occurrence of specific circumstances.

For these purposes, Art. 25, paragraph 5, of the OECD Model introduces a mandatory binding arbitration, should the competent authorities fail to reach an agreement pursuant to the mutual agreement procedure. However, arbitration shall not be an available option if a signatory State makes a reservation, so that the will of the State is still crucial for these proceedings to take place.

Under this light, arbitration agreements are extremely important in international agreements, where a general principle of good faith would suggest to the competent Fiscal Administration to avoid, to the maximum extent possible, any taxation that would violate such international agreements.

In this scenario, tax law analyses the application of arbitration proceedings to a different set of disputes, generally based on public law and occurring between taxpayers and the Fiscal Administration, as well as between different public bodies.

An inquiry from a tax law prospective is essential for a better and more far-reaching understanding of arbitration. All the more reason so why the European Union has recently adopted Directive n. 2017/1852/UE regarding double taxation dispute resolution mechanisms.

Directive n. 2017/1852/UE held that he elimination of double or multiple taxation was a necessary step, as they cause distortions and inefficiencies and have a negative impact on cross-border investment and growth. The mechanisms currently provided for in bilateral tax treaties, in fact, might not achieve the effective resolution of such disputes in an entirely desirable and timely manner, due to some important short-comings as to access to Mutual Agreement Procedures, length and cost-efficiency of proceedings.

The negative effects of inefficiency in dealing with double taxation affect the growth of cross-border transactions and has to confront itself with the more intensified and penetrating audit practices by tax administrations.

The aim of the Directive is to improve double taxation and transfer pricing dispute resolution mechanisms, as to ensure legal certainty providing a final decision of the case and eliminate double taxation. A great deal of attention is devoted to the improvement of the existing instrument in order to establish a reasonable time-limit for the duration of proceedings, a uniform scope of application within the European Union, as well as more efficiency and reliability in the enforcement of the final decision. A uniform application of these rules would immensely contribute to fight tax evasion and tax avoidance.

References

Azzali, S. & Tavartkiladze, B. (in press). Gli arbitrati secondo regolamenti precostituiti. In L. Dittrich (Ed.), *Trattato di procedura civile*. Turin: UTET.

Califano, C. (2004). Strumenti processuali per l'accesso del contribuente alla Corte Europea dei diritti dell'uomo. In F. Bilancia, C. Califano, L. Del Federico, & P. Puoti (Eds.), *Convenzione Europea dei diritti dell'uomo e Giustizia Tributaria Italiana* (pp. 435–449). Turin: Giappichelli.

Chizzini, A. (2015). Konvenzionalprozess e poteri delle parti. *Rivista Di Diritto Processuale, 91*, 45–60.

Di Stasi, A. (2013). *Ammortizzatori sociali e solidarietà post-industriale*. Torino: Giappichelli.

Di Stasi, A., & Torsello, L. (2015). Economic flourishing, social justice and legislation policies. *Scientific Journal of Economics and Business Management*. https://doi.org/10.15651/978-88-748-8868-9. 2/2 2015.

Mantucci, D. (2004). *Profili del contratto di subfornitura*. Naples: ESI.

Mantucci, D. (2008a). Nuove strategie di composizione dei conflitti endosocietari. In D. Mantucci (Ed.), *Conciliazione, arbitrato e gestione d'impresa nel nuovo diritto societario*. Naples: ESI. ss.; Id (2010) Nuove forme di tutela nei rapporti di credito al consumo. In G. Villanacci (Ed.), Naples: Credito al consumo.

Mantucci, D. (Ed.). (2008b). *Conciliazione, arbitrato e gestione d'impresa nel nuovo diritto societario*. Naples: ESI.

Perlingieri, P. (2006). *Il diritto civile nella legalità costituzionale secondo il sistema italo comunitario delle fonti* (3rd ed.). Naples: ESI.

Piperata, G. & Trucchia, L. (2016). Pubblico e privato nella scuola. In L. Ferrara & D. Sorace (Eds.), *A 150 anni dall'unificazione amministrativa italiana. Studi* (Vol. VI, pp. 525–559). Florence: Unità e pluralismo culturale.

Putti, P. (2014). La famiglia e la coppia in crisi nel caleidoscopio delle nuove tecniche di risoluzione delle controversie alternative alla giurisdizione. In M. De Angelis (Ed.), *Disponibilità delle situazioni soggettive e giustizia alternativa*, D.M., (pp. 87–114),

Quaranta, A. (2013). *La Tutela dei diritti fondamentali nella Costituzione*. Naples: ESI.

Tavartkiladze, B. (in press). Res Judicata and Issue Estoppel in enforcement proceedings under the New York Convention. *Journal of International Arbitration*.

Torsello, L. (2015). *Tutela del lavoro e procedure concorsuali*. Turin: Giappichelli.

Trucchia, L. (2018). Corte costituzionale e riparto di competenze tra Stato e Regioni in materia di pubblico impiego. *Il lavoro nelle pubbliche amministrazioni, 20*(1), 129–137.

Zarro, M. (2017). La tutela della reputazione digitale quale *intangible assets* dell'impresa. *Rassegna Di Diritto Civile, 37*(4), 1504–1527.

Zarro, M. (2018). Notazioni in tema di possesso degli *intangibles*: il caso del *know how*. *Il Foro nap., 120*(1), 183–204.

Zuccarino, S. (2014). L'arbitrato nelle controversie relative ai servizi di pubblica utilità: Energia elettrica e gas. In M. De Angelis (Ed.), *Disponibilità delle situazioni soggettive e giustizia alternativa*, D.M., (pp. 115–132).

From GDP to BES: The Evolution of Well-Being Measurement

Francesco Maria Chelli, Mariateresa Ciommi and Marco Gallegati

Abstract Measures of economic well-being are strictly linked to phases of economic development. While GDP may be considered a good indicator of well-being in the early phases of economic development, its relevance tends to progressively decrease in favor of alternative measures as countries go on the development path. Economists and Statisticians of the Università Politecnica delle Marche have always been at the forefront of this research agenda, thanks to the foresight of their master Giorgio Fuà. Inspired by his works, the Ancona's team has continued on his track, actively contributing to the debate on beyond GDP and suggesting alternative measures. Moreover, following Fuà's thought, that is, the relevance of the local dimension to evaluate the well-being of citizens, Ancona's scholars propose the measurement of well-being in Italy at National, Regional and local level, in the long and short run, through the construction of composite indicators as well as applying new statistical techniques.

1 Introduction

In the last years, there has been an increasing interest in measuring well-being. Scholars are now convinced that a unique monetary measure such as the Gross domestic product (GDP), is not enough to account for a multifaceted phenomenon such as individual well-being. Thus, it is necessary to add to the traditional measures additional monetary and non-monetary indicators to have a complete picture of the

F. M. Chelli · M. Ciommi · M. Gallegati (✉)
Department of Economics and Social Sciences,
Università Politecnica delle Marche, Piazzale Martelli 8, 60121 Ancona, Italy
e-mail: marco.gallegati@univpm.it

F. M. Chelli
e-mail: f.chelli@univpm.it

M. Ciommi
e-mail: m.ciommi@univpm.it

well-being of a country. To tell the truth, since its formulation, by Simon Kuznets in 1934,[1] the GDP has shown several limitations.[2]

By construction, the GDP provides a measure of economic activities which can be priced. Its link with the evaluation of well-being is a (more) recent interpretation. Anyway, trying to measure well-being through the GDP equals trying to extend the market principles to social life or, similarly, trying to measure space-time by liters or culture by kilos. That is, looking for an answer to a poorly asked question.

That the GDP was inadequate to account for well-being is commonly known. In fact, more then 50 years ago, Simon Kuznets warned that: *The welfare of a nation can scarcely be inferred from a measurement of national income... Especially when the economy is developed and goals for more growth should specify of what and for what i.e. the quantity from the quality of the growth and the long term consequences.*[3] In fact, when computing the GDP, for instance, we ignore the effects and damages that economic growth produces on the environment. That is, the GDP calculus does not take into account all these services that are without a market price, which means they are not tradable, but end up in a net loss for future (and present) generations. These services, such as the regulation of climate and atmospheric gases, the decomposition and absorption of waste, soil formation or pollination, just to mention a few, are invaluable and in no way could be replaced if they were degraded or irreversibly destroyed. Yet we distractedly observe the loss of biodiversity, deforestation and the consequent impact these have on the soil (erosion, slope instability, desertification, etc.), the atmosphere (on climate regulation at different levels), and on human communities (mass migration due to desertification).

In February 2008, the President of the French Republic Nicholas Sarkozy invited Joseph Stiglitz, Amartya Sen and Jean Paul Fitoussi to establish a Commission, the Commission on the Measurement of Economic Performance and Social Progress (CMEPSP), with the aim of identifying the limits of GDP as an indicator of economic performance and social progress, including the problems with its measurement.

In fact, while GDP may be considered a good indicator of well-being in the early phases of economic development, its relevance tends to progressively decrease in favor of alternative measures as countries go on the development path. In addition, different people have a different idea of what matters and, consequently, what is important for one country is not necessarily important for another one. However, even if international comparison is fundamental for evaluating relative positions of countries over time, the local (regional or provincial) dimension should be equally important. In fact, there are many advantages in calculating well-being indicators at sub-national local. Firstly, at a local level, it is possible to better capture territorial disparities and inequalities, that, at national level could be hidden. Moreover, local evaluations could help policy makers to address more efficient policies for their territories, better than a national measure. Jumping the gun, more than 25 years

---

[1] National Income, 1929–1932, Senate document no. 124, 73d Congress, 2d session.

[2] See Ciommi et al. (2013).

[3] The New Republic, October 20, 1962, cited in Cobb, Clifford, Ted Halstead, and Jonathan Rowe, *If the GDP is Up, Why is America Down?* The Atlantic Monthly, October 1995, page 67).

before the constitution of the Stiglitz-Sen-Fitoussi Commission and the publication of their results (see CMEPSP 2009; Stiglitz et al. 2009). Giorgio Fuà wrote:

> We ought to combat the dominant conception whereby one single model of development and of its life cycle (the model centered on the growth of marketable goods) is proposed and accepted as the only valid one. We ought to urge every population to seek the form of progress that matches better its history, its characteristics, its circumstances, and not to feel inferior merely because another country produces more goods. Today, although this may seem pure Utopia, one must nevertheless think about it.[4] (Fuà 1993)

Inspired by his work, Economists and Statisticians at the Università Politecnica delle Marche have investigated in-depth the relationship between the GDP and well-being, actively contributing to the debate on beyond GDP and proposing alternative measures. First, a team of scholars led by Giorgio Fuà contributed to annual estimates of the Italian National Institute of Statistics (ISTAT) of historical national accounts for the period 1861–1956 by including estimates of the value added by sector at constant price (1938), implicit deflators by sector and use, and the creation of a comprehensive series on the capital stock from 1881 onwards (Ercolani 1969, 1981a). These estimates were later modified by Fuà and Gallegati (1993) through the construction and application of a Paasche annual chain index to the Italian GDP real product between 1861 and 1989, a methodology adopted ten years later by ISTAT.

More recently, motivated by the increasing interest in combining GDP with other measures, such as indicators of inequality and sustainability and following Fuà's thought, that is, the relevance of the local dimension to evaluate the well-being of citizens, the Ancona scholars have contributed to the debate on beyond GDP. They have investigated these themes in-depth, adopting a multidimensional perspective of the Italian well-being measurement and proposing the measurement of well-being in Italy at National, Regional and local level, in the long and short run, through the construction of composite indicators as well as applying new statistical techniques. Their research agenda moves in parallel with the project promoted by ISTAT and CNEL, the so-called Equitable and Sustainable Well-being (BES)[5] aiming at proposing a dashboard of indicators to evaluate the progress of the Italian society not only from an economic, but also from a social and environmental perspective.

The rest of the paper is structured as follows. Section 2 takes a look at the past, retracing the steps that led to the measurement of economic growth in Italy after the Unity. Section 3 proposes a modern view of well-being, reporting what has been done by the Ancona team to measure well-being at sub-national level in recent years. Section 4 traces the lines of future research and Sect. 5 concludes.

---

[4] Authors' translation based on Fuà (1993).

[5] In fact, since 2013, ISTAT has annually published a Report on Equitable and Sustainable Well-being that provides an overview of the well-being of Italy and its Regions. See, ISTAT (2017) for more details.

2 Measurements of Economic Growth in Italy After Unity

Italy has a solid and long tradition of studies dealing with the issue of reconstructing national accounts in the post-Unification period. Estimates of Italy's National Accounts for the period after the Unity were first produced by the National Institute for Statistics (ISTAT) in ISTAT (1957). This set of historical national accounts for the period 1861–1956 included a detailed reconstruction of both the production and expenditure side at current prices, and also the estimate of the national product measured at 1938 constant prices, disaggregated with respect to using expenditure alone (obtained as sum of the expenditure components). However, details of the methods used and information on the sources of aggregate data, as well as a set of key historical series such as output by sector at constant prices, were missing.

A decade later a research team of young economists under the direction of Giorgio Fuà provided historical series at constant prices of national product disaggregated with regard to the formation of resources as well. The team's contributions included the estimates of the value added by sector at 1938 constant prices (Vitali 1969), and the implicit deflators by sectors and use, obtained dividing the ISTAT series at current prices by the Vitali series (Ercolani 1969). The ISTAT-Fuà GDP series has been used as reference in all subsequent revisions of national accounts. Maddison (1991) presented a new estimate of Italy's GDP series for the period 1861–1989 using ISTAT-Fuà series for agriculture and services and industrial series by Fenoaltea (1988), (that lowered the initial levels of the ISTAT-Fuà GDP series) and increased the overall growth rate of Italy's GDP between 1861 and 1913. Then Rossi et al. (1993) reconstructed the GDP series from the expenditure side for the period 1890–1990 by revising the benchmark of 1911, 1938 and 1951 (the first provided by the Bank of Italy, Rey 1991, the last by Golinelli and Monterastelli 1990) and superimposing the old cyclical component of the ISTAT-Fuà series. Finally, Baffigi (2013) provided a new GDP series, together with supply and demand side estimates, covering the 150 years after Italy's political Unification, as part of the *150 years* national accounts project[6] that includes the Bank of Italy, ISTAT, as well as researchers from other institutions. All in all, these series are dominated by ISTAT components, and apart from a trend correction, their path remain extremely close to that of the ISTAT-Fuà aggregate.[7]

---

[6]The project aims at reconstructing long series of several Italian economic and non-economic variables to celebrate the 150 years since Italy's political Unification. See http://www.bancaditalia.it/pubblicazioni/quaderni-storia/index.html.

[7]The new estimates of GDP (Baffigi 2013) makes exception to this rule. There is evidence of a more regular pattern of the Italian economic development in the 1880–1910 period that suggests a different alternative interpretation of the phases of Italy long-run economic growth, especially regarding the interpretation that traces back Italy's economic development to the beginning of the nineteenth century, i.e. the Giolittian era, (e.g. Fuà 1981; Maddison 1991; Rossi et al. 1993). Indeed, the first forty years of life of the newly born Italian Kingdom are now viewed as a phase of slight but significant growth, not as stagnant decades followed by a sudden and sharp acceleration in the pace of growth around the end of the nineteenth century.

ISTAT estimates at constant prices, as revised by Fuà and his group and known as ISTAT-Fuà estimates, has represented the reference series for detecting the long-term growth rates of the Italian economy and identifying growth-phase periods of the Italian economic development, e.g. Fuà (1969, 1981), Ciocca and Toniolo (1796), Toniolo (1978, 1988), Zamagni (1993), Ciocca (1994). Moreover, the cyclical component of the ISTAT-Fuà series, isolated through the trend-cycle decomposition method, has been the object of studies, especially in recent years, on the nature and causes of business cycle fluctuations. See, for instance, Ardeni and Gallegati (1991, 1994), Gallegati and Stanca (1998), Delli Gatti et al. (2005) and Clementi et al. (2015).

The expertise accumulated on the techniques of estimation of aggregate national accounting raised Fuà's awareness about two main drawbacks on the use of statistical measures of growth at constant prices for comparison purposes. The first is a measurement problem related to using a single system of constant prices for intertemporal comparisons of income for the entire time-span. To overcome this problem more sophisticated weighted procedures were later introduced by Fuà and Gallegati (1993). The authors proposed to replace the constant-price reduction method with the chain method for the construction of an annual chain index of Italy's real product for the period 1861–1989, a technique that is now standard for the estimation of GDP at constant prices. The latter is to what extent GDP may be used as a valid indicator of a population's welfare (well-being). Several indicators, like life expectancy at birth, suggest a divergence between income and well-being along the developmental path. Economic growth may be a useful indicator of well-being at the early stages of economic development of a modern economy, but for advanced mature economies a set of (better) alternative indicators needs to be developed (Fuà 1993).

3 A Modern View: The Equitable and Sustainable Well-Being at Sub-national Level

The measurement and monitoring of economic performance before, and well-being in the recent years, have a long tradition. Figure 1 reports a brief evolution of such measurement, citing some of the most distinguished phrases that have led to the the modern idea of well-being. The measurement moves from indicators concerning the only economic dimensions (material stage) to indicators that account economic and social dimension simultaneously (social stage) up to embrace the environmental aspects (global stage).

At the end of XX century, there was an increasing interest in beyond GDP measurement. After the Sen-Fitoussi-Stiglitz Report (Stiglitz et al. 2009), experts such as economists, statisticians, as well as politicians and common people, have been working towards constructing and defining new multidimensional indicators and criticizing the existing ones.

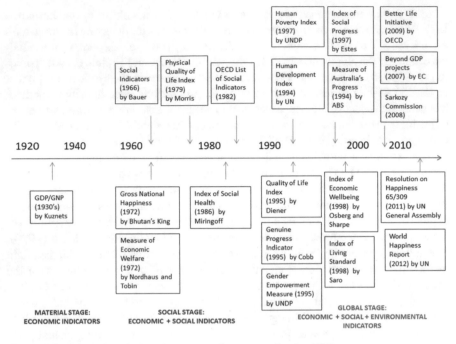

Fig. 1 Chronological evolution of measures of progress related to GDP

Under the *Beyond GDP initiative* and motivate by the above mentioned results, several national statistical offices, researchers as well as agencies proposed their own well-being measure as alternative or complementary measure to GDP.[8]

Also for Italy, several indexes have been computed. For instance, Gigliarano et al. (2014) provide a time series evaluation of the Italian well-being by proposing an alternative definition of the *Index of Sustainable Economic Welfare* (ISEW) based on a weighting scheme for private consumptions according to a poverty measure. They compute the index for Italy and for all its regions and macro-areas, providing also a ranking of the Italian macro-areas and regions based on ISEW in comparison with the ranking based on GDP. The analysis is annual and covers the decade 1999-2009. This is a first attempt to provide a time series of the ISEW in the long run and at this disaggregated level.

In Italy, the Beyond GDP initiative has given rise to the Equitable and Sustainable Well-Being (BES) project,[9] coordinated by the National Council for Economics and Labor (CNEL) and by the Italian National Institute of Statistics (ISTAT). The

---

[8]See http://ec.europa.eu/environment/beyond_gdp/index_en.html for more information on the *Beyond GDP initiative* and Ciommi et al. (2013) for a detailed description of the most relevant indices proposed.

[9]*Benessere Equo e Sostenibile* (in Italian). See https://www4.istat.it/en/well-being-and-sustainability/well-being-measures/bes-report for more details.

BES aims at complementing GDP, by taking into account social and environmental information as well as measures of inequality and sustainability. The BES is thought of as a dashboard of 134 indicators partitioned into 12 domains of well-being.[10] The indicators are computed at national and regional (NUTS2) level. Even if in the original formulation, the BES was conceived as a dashboard of indicators, since 2015 a first synthetic measure for the 12 domains has been proposed.

Inspired by the definition of this new dashboard and interested in the long run measurement, the Ancona researchers began to study the temporal trend of multidimensional well-being of the Italians from 1861 to 2011 by country and regions (see Ciommi et al. 2017; Chelli et al. 2018, respectively). On the one hand, they annually computed the BES for Italy by constructing 41 indicators[11] grouped into eight dimensions (health, education, work, economic well-being, political participation, security, environment and research and development) and computing a composite index for each domain. The analysis of their trends over time provide an interesting picture of Italian history and reveal scenarios that would remain hidden focusing only on the GDP. On the other hand, they analyse the well-being census at regional level. Collecting 21 indicators partitioned into six well-being dimensions, they take a picture of the territorial gap of Italian regional development by means of a cluster analysis. They find that the only North-South division is not sufficient to frame the situation, it would be an incomplete configuration both with regard to the historical context and in relation to the examined dimensions of the development.

Motivated by the increasing interest in the local measurement of well-being, besides the national and regional computation, in 2011 the Pesaro-Urbino Province, under ISTAT's scientific and technical supervision, promotes the measurement of the BES at provincial level (NUTS3), the so-called *Provinces' BES Project*. In 2018, a first complete database for all 110 Italian provinces, consisting of 61 and 11 domains was launched.[12]

Compared to the *Provinces' BES Project*, since 2016, researchers from Ancona have provided exploratory analysis on a preliminary dataset produced by ISTAT consisting of 88 indicators for all the Italian provinces in 2014 (Chelli et al. 2016; Ciommi et al. 2017b). The research moved in two distinct directions: the application of statistical techniques to trace the elementary indicators and construct a summary index for each domain and the methodological proposal of new indicators. On the one hand (Chelli et al. 2016) after having identified 70 indicators partitioned in 11 domains, they propose to reduce the redundant information within each BES domain by applying a factor analysis and take into account the distribution of these factors across the provinces by applying a cluster analysis. This allows to identify which groups of provinces display similar behavior according to the different well-being

---

[10]Namely Health, Education and training, Work and life balance, Economic well-being, Social relationships, Politics and Institutions, Safety, Subjective well-being, Landscape and cultural heritage, Environment, Research and innovation and Quality of services.

[11]The selection depended on three factors: (*i*) maximum intersection with the original BES; (*ii*) maximum overlapping with similar international proposals; (*iii*) availability of the time series.

[12]See http://www.besdelleprovince.it/ for more details.

dimensions. On the other hand (Ciommi et al. 2017b), the authors, after selecting the 41 indicators according to reliability and robustness criteria, propose a parametric form for composite indicators based on two parameters proving that different choices of such parameters induce to six different aggregation methods (the simple arithmetic mean, the Adjusted Mazziotta-Pareto Index, and their modifications based on weighting systems depending on the Gini coefficient and on its reciprocal).

4 What the Future Holds

In Italy, the debate on going beyond GDP and measuring well-being and social progress is still restless. On the one hand, several variables belonging to the BES are now in the Italian Economic and Financial Document.[13] Consequently, the measurement of well-being through the BES variables is crucial, especially its measurement at local level and, more precisely, at municipality level. For this reason, new research is now moving in that direction. With the recent public release of municipal data by ISTAT, new research scenarios are opening in the analysis of data at the local level and in the construction of composite indicators. Therefore, multivariate spatial statistical techniques will be combined with traditional aggregative techniques, with the aim of better capturing local disparities and obtain timely assessments of the effectiveness of public policies. However, *Historia magistra vitae est*[14] and, consequently, we can not ignore the study of the past since it should serve as a lesson to the future. Thus, research should also be devoted to better understanding local well-being from an historical perspective, trying to better measure it at regional and local levels. Indeed, today's strong economic gap between north and south Italy can be partly understood by historical analysis. Therefore, the analysis of the regional well-being dimensions, besides the economic one, could help to uncover these differences.

On the other hand, in September 2015, the United Nations General Assembly adopted the so-called *Agenda 2030* (Transforming our world: the 2030 Agenda for Sustainable Development [15]). The *Agenda* traces the path that each country has to travel over the next 15 years with the aim of ending poverty, protecting the planet and ensuring prosperity for all. The document consists of 169 targets grouped into 17 objectives, called Sustainable Development Goals (SDGs) and on March 2016, Inter Agency Expert Group on SDGs (IAEG-SDGs) proposed a first list of 241 indicators to identify a shared statistical information framework aiming at monitoring and evaluating progress towards the objectives of the Agenda. Therefore, a connection

---

[13]The budget reform demands that the BES plays a crucial role in the process of defining economic policies by accounting on the effects of such policy also on some fundamental dimensions that quantify the quality of life. Law n. 163/2016 of the Italian government. See http://www.mef.gov.it/inevidenza/article_0276.html for more details.

[14]This is a Latin expression, taken from Cicero's *De Oratore (II, 369)*. The English translation is "history is life's teacher".

[15]http://www.un.org/en/development/desa/population/migration/generalassembly/docs/globalcompact/A_RES_70_1_E.pdf.

between BES and SDGs is indispensable, especially with a view to ensuring international comparisons. Keeping always in mind the importance of the local well-being, an effort in the measurement of SDGs at local level is also required.

Finally, it is time to compute these well-being indicators for population subgroups. The idea of well-being of a young person is completely different from that of a older one. What really matters for the two groups is not identical. The same applies to women and men. Consequently, well-being indicators, and in particular the indicator at local level, should be able to capture these differences.

So it is the researchers' task to formulate well-being indices that are decomposable by subgroups from the statistical and mathematical point of view.

5 Conclusions

Never like today, does Fuá's teaching seems to be more contemporary. The monetary variable (income) alone is not able to grasp the well-being of a country like Italy.

> What sense is it to rejoice in the growth of a product if it turns out that pollution has increased and it has become dangerous to walk down the street? In cases like these our well-being gets worse. The same applies to inflation that takes into account how much we pay for goods, but not, for example, the quality of life.[16] (Fuà 1994)

The main purpose of any Government should be the implementation of public policy with an objective of increasing overall well-being of its citizens. Well-being understood as the peoples ability to live the lives they value. If it is true that the purpose of public policy is to improve the well-being of individuals and communities, then its measurement should be of public interest.

The Ancona team is actively working in this direction, building on past teachings and hoping to continue to be inspirational for young researchers.

References

Ardeni, P. G., & Mauro, Gallegati. (1991). Long-term trends and cycles in the Italian economy (1861–1988). *Giornale degli Economisti e Annali di Economia, 50*, 193–235.

Ardeni, P. G., & Mauro, Gallegati. (1994). Crescita e fluttuazioni nell'economia italiana (1861–1913). *Una reinterpretazione, Politica Economica, 10*, 119–153.

Baffigi, A. (2013). National accounts, 1861—2011. In G. Toniolo (Ed.), *The Oxford handbook of the Italian economy since unification* (pp. 157–186). New York: Oxford University Press.

BES. (2017). ISTAT. *Il Benessere Equo e sostenibile in Italia*. Roma. https://www.istat.it/it/files/2017/12/Bes_2017.pdf.

Clementi, F., Gallegat, M., & Gallegati, M. (2015). Growth and cycles of the Italian economy since 1861: The new evidence. *Italian Economic Journal, 1*(1), 25–59.

---

[16]Authors' translation based on Fuà (1993).

Chelli, F. M., Ciommi, M., Emili, A., Gigliarano, C., & Taralli, S. (2016). Assessing the equitable and sustainable well-being of the Italian provinces. *International Journal of Uncertainty, Fuzziness and Knowledge-Based Systems*, *24*(Suppl. 1), 39–62.

Chelli, F. M., Ciommi, M., Ermini, B., Gallegati, M., Gentili, A., & Gigliarano, C. (2018). San Matteo e la provvidenza. I luoghi ei tempi dello sviluppo italiano. *Rivista giuridica del Mezzogiorno*, *32*(3), 643–672.

Ciocca, P. (Ed.). (1994). *Il progresso economico dell'Italia. Permanenze, discontinuità*. Bologna: limiti, Il Mulino.

Ciocca, P., & Toniolo, G. (Eds.). (1976). *L'economia italiana nel periodo fascista*. Bologna: Il Mulino.

Ciommi, M., Gigliarano, C., Chelli, F. M., & Gallegati, M. (2013). Behind, beside and beyond GDP: Alternative to GDP and to macro-indicators. *E-Frame EU-Project. Deliverable 3*(1). http://www.eframeproject.eu/fileadmin/Deliverables/Deliverable3.1.pdf.

Ciommi, M., Gentili, A., Ermini, B., Gigliarano, C., Chelli, F. M., & Mauro, Gallegati. (2017a). Have your cake and eat it too: The well-being of the Italians (1861–2011). *Social Indicators Research*, *134*(2), 473–509.

Ciommi, M., Gigliarano, C., Emili, A., Taralli, S., & Chelli, F. M. (2017b). A new class of composite indicators for measuring well-being at the local level: An application to the equitable and sustainable well-being (BES) of the Italian provinces. *Ecological Indicators*, *76*, 281–296.

CMEPSP. (2009). Survey of existing approaches to measuring socio-economic progress. *Commission on the measurement of economic performance and social progress*. https://www.insee.fr/en/information/2662494

Delli Gatti, D., Gallegati, M., & Gallegati, M. (2005). On the nature and causes of business fluctuations in Italy, 1861–2000. *Explorations in Economic History*, *42*, 81–100.

Ercolani, P. (1969). Documentazione statistica di base. In G. Fuà (Ed.), *Lo sviluppo economico in Italia. Storia dell'Economia Italiana negli ultimi cento anni. Volume III, Studi di settore e documentazione di base* (pp. 380–460). Milano: Franco Angeli.

Fenoaltea, S. (1988). The extractive industries in Italy, 1861–1913: General methods and specific estimates. *Journal of European Economic History*, *17*(1), 117–125.

Fuà, G. (1966). *Notes on Italian economic growth 1861–1964*. Milano: Ciaffré.

Fuà, G. (Ed.). (1969). *Lo sviluppo economico in Italia. Storia dell'Economia negli ultimi cento anni. Volume III—Studi di settore e documentazione di base*. Milano: Franco Angeli.

Fuà, G. (Ed.). (1981). *Lo sviluppo economico in Italia. Storia dell'Economia Italiana negli ultimi cento anni. Volume I—Lavoro e reddito*. Milano: Franco Angeli.

Fuà, G. (1993). *Crescita economica. Le insidie delle cifre*. Bologna: Societa editrice Il Mulino.

Fuà, G., & Gallegati, M. (1993). An annual chain index of Italy's "Real" product, 1861–1989. *Review of Income and Wealth*, *42*(2), 207–224.

Fuà, G. (1994). Crescita, benessere e compiti dell'economia politica. *Il Mulino, Rivista bimestrale di cultura e di politica*, *43*(5), 761–768.

Gallegati, M., & Stanca, L. M. (1998). *Le fluttuazioni economiche in Italia, 1861–1995. Ovvero, il camaleonte e il virus dell'influenza*. Torino: Giappichelli.

Gigliarano, C., Balducci, F., Ciommi, M., & Chelli, F. M. (2014). Going regional: An index of sustainable economic welfare for Italy. *Computers, Environment and Urban Systems*, *45*, 63–77.

Golinelli, R., & Monterastelli, M. (1990). *Un metodo di lavoro per la ricostruzione di serie storiche compatibili con la nuova contabilità nazionale (1951–1989), Nota di lavoro n. 9001*. Bologna: Prometeia.

ISTAT. (1957). *Indagine statistica sullo sviluppo del reddito nazionale in Italia dal 1861 al 1956, Annali di statistica, serie VIII* (Vol. IX). Roma.

Maddison, A. (1991). A revised estimate of Italian economic growth, 1861–1989. *BNL Quarterly Review*, *177*, 225–241.

Rey, G. M. (Ed.). (1991). *I conti economici dell'Italia. I. Una sintesi delle fonti ufficiali. 1890–1970*. Roma-Bari: Laterza.

Rossi, N., Sorgato, A., & Toniolo, G. (1993). I conti economici italiani: una ricostruzione statistica, 1890–1990. *Rivista di Storia Economica, 10*, 1–47.

Stiglitz, J. E., Sen, A., & Fitoussi, J.-P. (2009). *Report by the commission on the measurement of economic performance and social progress.* https://ec.europa.eu/eurostat/documents/118025/118123/Fitoussi+Commission+report. Accessed November 15, 2018.

Toniolo, G. (Ed.). (1978). *L'economia italiana 1861–1940.* Bari: Laterza.

Toniolo, G. (1988). *Storia economica dell'Italia liberale (1850–1918).* Bologna: Il Mulino.

Vitali, O. (1969). La stima del valore aggiunto a prezzi costanti per rami di attività. In G. Fuà (Ed.), *Lo sviluppo economico in Italia* (Vol. 3). Milano: Franco Angeli.

Zamagni, V. (1993). *The economic history of Italy, 1860–1990: Recovery after decline.* Oxford: Clarendon Press.

Quantitative Methods in Economics and Finance

Luca Vincenzo Ballestra, Serena Brianzoni, Renato Colucci, Luca Guerrini, Graziella Pacelli and Davide Radi

Abstract *Evaluation of Financial and Actuarial Risk.* The research group has focused on several different aspects of quantitative finance. As a first research subject, we faced the problem of pricing credit default swaps (CDSs), which entails the calculation of the risk of default. As a second research subject, we considered the problem of pricing complex derivatives. Precisely, we took into account barrier options on an underlying described by either the geometric fractional Brownian motion or a time-changed Brownian motion. Finally, we dealt with the problem of pricing real options in the presence of stochastic interest rates. *Nonlinear Dynamics in Economic and Financial Models.* Nonlinear Dynamics is an interdisciplinary area characterized by a rapid and extensive development in recent years, which has proved to be very useful in explaining some important facts in Economics and Finance (such as endogenous fluctuations). In this contribution, we provide an overview of our research on this topic, both in continuous and in discrete time.

L. V. Ballestra
Department of Statistical Sciences,
Università Alma Mater Studiorum Bologna, Bologna, Italy
e-mail: luca.ballestra@unibo.it

S. Brianzoni · L. Guerrini · G. Pacelli (✉)
Department of Management, Università Politecnica delle Marche, Ancona, Italy
e-mail: g.pacelli@univpm.it

S. Brianzoni
e-mail: s.brianzoni@univpm.it

L. Guerrini
e-mail: luca.guerrini@univpm.it

R. Colucci
Department of Engineering, Università Niccolò Cusano, Roma, Italy
e-mail: renato.colucci@unicusano.it

D. Radi
Department of Economics and Management, Università di Pisa, Pisa, Italy
e-mail: davide.radi@unipi.it

© Springer Nature Switzerland AG 2019
S. Longhi et al. (eds.), *The First Outstanding 50 Years of "Università Politecnica delle Marche"*,
https://doi.org/10.1007/978-3-030-33879-4_9

117

1   Evaluation of Financial and Actuarial Risk[1]

The activity of the research group in Mathematical Methods for Economics, Actuarial Science and Finance of the University of Ancona focuses on several different aspects of quantitative finance. Special emphasis is put on the development of new theoretical models and new numerical and analytical methods for computing the default risk, for pricing financial derivatives and for conducting empirical tests to validate the models. These contributions are on one hand theoretically relevant and on the other hand useful for financial practitioners. In the following, we present and discuss the main topics covered by our researches.

A relevant issue tackled in our researches is the assessment of the risk of default of companies. Pricing the risk of default is a fundamental task for several financial market players such as corporate bond investors, credit derivative traders, banks, mortgage suppliers and insurance companies. To this aim, various models of credit risk have been developed which are based on two different approaches: the structural approach and the reduced-form approach.

Structural models describe the default event by means of one or more variables related to the capital structure of the firm issuing the debt. For example, according to the first proposed structural model, which has been developed in Merton (1974), a firm defaults if at the debt maturity the value of its assets is lower than the value of its obligations. An improvement of this model is presented in Black and Cox (1976), where the possible occurrence of premature bankruptcy as well as the debt seniority are taken into account. Other more sophisticated structural models are also available which incorporate variables such as the interest rate Longstaff and Schwartz (1995), Briys and de Varenne (1997), Bernard et al. (2005), tax benefits Anderson and Sundaresan (1996), debt restructuring Abínzano et al. (2009), liquidation costs Leland and Toft (1996) and downgrade-triggered termination clauses Feng and Volkmer (2012). As revealed by several empirical studies, see, e.g, Jones et al. (1984), Franks and Torous (1989), such a kind of models have a major issue: if the firm's assets value is specified as a continuous-time stochastic process, then for short debt maturities the probability of default that the models predict turns out to be very close to zero, contrary to what happens in reality, see, e.g, Crouhy et al. (2000), Bäuerle (2002). In order to account for high short-term spreads, some authors, see, e.g. Zhou (2001), Chen and Panjer (2003), have proposed structural models with unexpected jumps in the firm asset value. Nevertheless, this approach lacks analytical tractability, which makes it difficult to calibrate the model parameters to observed credit spreads.

To overcome the issues of the structural models, reduced form models have been proposed. According to the reduced-form approach, see, e.g, Duffee (1999), Duffie and Singleton (1999), Madan and Schoutens (2008), Schoutens and Cariboni (2009), Fontana and Montes (2014) the default event is modeled as the first jump of a counting process whose intensity, termed intensity of default, is not assumed to be firm-specific but is prescribed exogenously. This allows one to take into account the possible occurrence of a sudden (unpredictable) default event and henceforth the high credit

---

[1]*L.V. Ballestra, G. Pacelli, D. Radi.*

spreads that are often experienced for short debt maturities can be recovered. In addition, reduced-form models are relatively simple from a mathematical standpoint and thus they usually offer a large amount of analytical tractability.

However, reduced-form models have the heavy drawback of not taking into account any information about the capital structure of the firm. Such an issue has prompted some authors to develop hybrid models of credit risk in which the reduced-form approach is combined with some structural variable, see, e.g, Madan and Unal (1998), Madan and Unal (2000), Duffie and Lando (2001), Cathcart and El-Jahel (2003), Cathcart and El-Jahel (2006), Giesecke (2006), Ballestra and Pacelli (2014).

Among these models, the one proposed in Madan and Unal (1998) deserves a special attention as it is a parsimonious hybrid model. In particular, it is mainly developed based on the reduced-form approach, but the default intensity, instead of being prescribed exogenously, is specified as a convenient function of the firm's equity value. This allows us to recover the desirable features of both the structural and the reduced-form models, but at the same time the parameters involved are only one more than the parameters of Merton's model, see Merton (1974). Therefore, the approach by Madan and Unal turns out to be particularly appealing and suitable for practical uses.

The model in Madan and Unal (1998) does not have an analytical closed-form solution. To be more precise, Madan and Unal have provided a closed-form expression for the survival probability, but, as pointed out in Grundke and Riedel (2004), the procedure used to derive this formula is not mathematically correct. Consequently, as shown in Grundke and Riedel (2004), where the default probability is computed by finite difference approximation, the closed-form solution obtained by Madan and Unal yields a survival probability that can also differ substantially from the true survival probability of the model. In order to compensate for the lack of a closed-form solution, we developed two alternative methods to approximate it.

The first one is proposed in Ballestra and Pacelli (2009), and it is an analytical approximation of the survival probability of the model by Madan and Unal. This formula is fairly accurate and computationally fast, but it is applicable only if one of the model parameters is sufficiently small (as it is based on a perturbation approach, see Ballestra and Pacelli 2009).

To provide a more general approximation of the survival probability in the Madan and Unal framework, which can also be useful for more general purposes, e.g. pricing financial derivatives such as credit default swaps (CDS, hereafter), a second method is proposed by Ballestra et al. (2017). It conducts to a quasi-analytical approximation of the survival probability in the model by Madan and Unal (1998). Such a formula, which is based on a Laplace-transform approach, turns out to be very accurate and computationally fast. Remarkably, this analytical expression for the survival probability allows us to price credit default swaps (CDSs) very easily. Specifically, a quasi-analytical formula to compute CDS par spreads is derived which is used in the aforementioned manuscript to calibrate the Madan-Unal credit risk model by fitting realized CDS par spreads. In particular, CDS names with different Moody's ratings are considered and the agreement between theoretical and empirical data is rather satisfactory, especially if we think that the stochastic differential equations on which the model by Madan and Unal stands involve only two unknown parameters. The pricing of CDSs has also been considered in Andreoli et al. (2016a), where the risk of default is specified according to a reduced-from model in high dimension. In particular, the interest rate and the survival probability are modeled by using a system of stochastic differential equations with either four or six factors, which generalizes an earlier model developed in Andreoli et al. (2015) and leads to a partial differential equation that is solved by means of a high-order finite difference scheme. Finally, a method to value CDOs, i.e. large portfolios of credit instruments, has been proposed in Andreoli et al. (2016b). Specifically, to deal with the incompleteness of the CDO market (due by the non-tradability of the survival probability), a Sharpe ratio approach is developed which generalizes the actuarial model presented in Bayraktar and Young (2007), Young (2008) and allows one to take into account non-hedgeable risk as well. This technique yields a set of partial differential equation in two independent space variables, which is efficiently solved by finite difference approximation.

Another relevant problem in quantitative finance is the empirical behavior of stock returns. In classical models for assessing the risk of default, stock returns are assumed to follow a geometric Brownian motion. This implies that many empirically observed features of the stock returns are neglected when assessing the solvency probability of a company. The neglected features of the logarithmic returns of the stocks include stylized facts such as self-similarity, heavy tails, long-range dependence and volatility clustering, see, e.g., Lo (1991), Ding et al. (1993) and Zhang et al. (2014). To overcome the mismatch between real stock returns and their theoretical representation, the fractional Brownian motion is often considered as an alternative formulation. It is a generalization of the Brownian motion that allows to replicate the stylized facts of the financial markets such as distribution of stock returns characterized by self-similarity, heavy tails, long-range dependence and volatility clustering, see, e.g., Lo (1991), Ding et al. (1993) and Zhang et al. (2014).

This modeling framework is however problematic for risk-neutral pricing as it does not allow one to construct a self-financing strategy yielding the risk-neutral price of financial options, see, e.g., Cheridito (2003) and Bender and Elliott (2004).

To include these striking features of the distribution of the stock returns in models for pricing financial derivatives under the risk-neutral approach, we introduce a mixed fractional Brownian motion (MFBM, hereafter) in a classical Merton modeling framework. The MFBM is a generalization of the fractional Brownian motion obtained as a linear combination of the fractional Brownian motion itself, see, e.g., Mandelbrot and Van Ness (1968), Duncan et al. (2000), Wang et al. (2001), Lim and Muniandy (2002, 2003), Longjin et al. (2010), Rostek and Schöbel (2013) and Hao et al. (2014), and of the standard Brownian motion, see, e.g., Øksendal (2003) (for other possible generalizations of the fractional Brownian motion the interested reader is referred to Wang et al. 2003, 2006, 2012; Liang et al. 2010; Gu et al. 2012). This stochastic process is particularly useful for our purposes because it allows to replicate the stylized facts of the distribution of the stock returns and, at the same time, it allows to construct a self-financing strategy. This makes it suitable for risk-neutral pricing.

In a recent contribution, see, e.g. Ballestra et al. (2016), we employed the MFBM to construct models for assessing the risk of default of a company and to price financial options. When pricing financial derivatives under the MFBM, the mathematical formulas to use are often more complicated than the classical ones developed assuming a Black-Scholes market. Moreover, some numerical approximations and techniques developed ad-hoc are often required. In this respect, we dealt with the problem of pricing barrier options on an underlying described by the mixed fractional Brownian model. To this aim, we considered the initial-boundary value partial differential problem that yields the option price and we derived an integral representation of it in which the integrand functions must be obtained solving Volterra equations of the first kind. In addition, we developed an ad-hoc numerical procedure to solve the integral equations obtained. Numerical simulations reveal that the proposed method is extremely accurate and fast, and performs significantly better than the finite difference method.

The models developed in finance are often used in management science to evaluate the opportunities of investments, the so-called real option analysis. Managers use real option analysis, also termed real option valuation, to decide about investment projects that can be undertaken at a future time, see, e.g., McDonald and Siegel (1986), Myers and Majd (1990), Dixit and Pindyck (1994), Charalampopoulos et al. (2001), Baldi and Trigeorgis (2009), Manley and Niquidet (2010), Fernandes et al. (2011b), Fernandes et al. (2011a), Bernardo et al. (2012), Santos et al. (2014), Nadarajah et al. (2015), Loureiro et al. (2015) and Nadarajah et al. (2017). Such an approach is particularly appealing because it allows one to take into account both the uncertainty and the flexibility related to the project to be valued. In this respect, we investigate in a recent contribution, see Ballestra et al. (2014), the problem of assessing investment projects under non-constant interest rates, see also Ingersoll and Ross (1992), Alvarez and Koskela (2006) and Schulmerich (2010). To this aim, a mathematical model is proposed where the revenue generated by investment projects is modeled as a geometric Brownian motion, and the interest rate is specified as a stochastic differential equation of Vasicek type Vasicek (1977). Under these assumptions, the problem of assessing the value of an investment opportunity is similar to the problem of pricing European vanilla options which can be solved in closed-form, see, e.g. Rabinovitch (1989).

The empirical analysis is done by considering several companies belonging to different production sectors and operating in the euro area. The results obtained show that investment projects are overvalued if the interest rate is assumed to be constant rather than stochastic. Nevertheless, the interest rate uncertainty does not have a substantial impact on the evaluation of firms' investments as confirmed by an empirical investigation underlying that the difference between the values of the investment projects under constant and stochastic interest rates is not statistically significant.

2 Nonlinear Dynamics in Economic and Financial Models[2]

It is well-known that economic and financial variables show fluctuations, even without exogenous shocks. In fact, in spite of the absence of external forces, the Economy can be unstable. For this reason, it is very useful to model complex economic systems as *Nonlinear Dynamical Systems*, which are able to explain endogenous fluctuations (see e.g. Dieci et al. 2014; Hommes 2013). In the following, we deal with Nonlinear Dynamical Systems both in discrete and continuous time, as natural words for studying important stylized facts in Economics and Finance.

2.1 Discrete Dynamical Systems

Many theoretical problems in Economics and Finance are formalized as dynamical systems in discrete time, also thanks to the large variety of methods and techniques (analytical and numerical) that this area offers. Concerning this subject, we shared in several researches, some of them are described in the following (Brianzoni et al. 2007, 2009, 2010a, b, c, 2011, 2012, 2015a, b, 2018).

2.1.1 Models with Differential Savings and Non Constant Population Growth Rate

A first line of research in discrete time concerns neoclassical one-sector growth models with differential savings, in order to investigate the possibility of complex dynamics. Moreover, for several production functions, the case of non constant population growth rate has been considered. In fact, when labor force is assumed to grow at a constant rate, population grows exponentially, which is clearly unrealistic given the carrying capacity of the environment.

The paper by Brianzoni, Mammana and Michetti, *Complex dynamics in the neoclassical growth model with differential savings and non-constant labor force growth*, analyzes the dynamics shown by the neoclassical one-sector growth model with differential savings, CES production function and the labour force dynamics described by the Beverton Holt equation. The resulting dynamical system is bidimensional, autonomous and triangular. The study of its qualitative and quantitative dynamic properties confirms that the system can exhibit cycles or even a chaotic dynamic pattern, if shareholders save more than workers, when the elasticity of substitution drops below one (so that capital income declines).

In order to take into account that the population growth rate can exhibit more complex dynamics, the paper by Brianzoni, Mammana and Michetti, *Non-linear dynamics in a business-cycle model with logistic population growth*, makes use of the logistic map. The resulting model is a bidimensional triangular dynamic system.

---

[2]*S. Brianzoni, R. Colucci, L. Guerrini.*

The study of its long run behaviour is developed as regards the parameter values. More precisely, the existence of the compact global attractor is proved. Moreover, the shape of the chaotic attractor changes in consequence of some global bifurcations, as some parameters vary. The analysis of these bifurcations is performed by using the critical curves method. The results confirm the central role of the production function's elasticity of substitution in the creation and propagation of complicated dynamics as in models with explicitly dynamic optimizing behaviour by private agents. Another parameter responsible for chaotic dynamics is the one in the logistic map, related to the amplitude of fluctuations in the population growth rate.

Differently, the paper by Brianzoni, Mammana and Michetti, *Variable elasticity of substitution in a discrete time Solow-Swan growth model with differential saving*, assumes that the technology is described by a variable elasticity of substitution (VES) production function. Differently from CES, VES production function allows to consider that the elasticity of substitution between capital and labor can be affected by a change in the level of the capital per-capita within the economic system and also the capital accumulation and output depend on such a change. Multiple equilibria are likely to emerge, according to the parameter values. Moreover, the model can exhibit unbounded endogenous growth when the elasticity of substitution between labour and capital is greater than one, as it is quite natural while the variable elasticity of substitution is assumed (and differently from CES). Being the final system defined by a continuous piecewise map, the study involves further mathematical methods in order to assess the possibility of complex dynamics (cycles of high periods or chaotic patterns) to be exhibited. In fact, a different type of bifurcations can occur, i.e. border-collision bifurcations, which are related to the contact of an invariant set with the border separating the regions of different definition of the map.

In order to generalize the previous results, the paper by Brianzoni, Mammana and Michetti, *Local and global dynamics in a neoclassical growth model with non concave production function and non constant population growth rate*, considers sigmoidal production function. The final two-dimensional dynamic system, describing the capital per capita and the population growth rate evolution, admits two coexisting attractors, whose structure becomes more complicated when the elasticity of substitution between production factors is low enough.

2.1.2   Growth Models with Corruption in Public Procurement

Another field of research focuses on the relationship between corruption in public procurement and economic growth within the Solow framework in discrete time. To be more precise, in the paper by Brianzoni, Coppier and Michetti, *Complex dynamics in a growth model with corruption in public procurement*, the public good is an input in the productive process and the State fixes a monitoring level on corruption. After solving a one-shot game via the backward induction method, a triangular piecewise smooth dynamic system is obtained. The resulting system admits multiple equilibria with nonconnected basins, moreover the existence of a global compact attractor is proved. The analysis of the local and global bifurcations (such as border collision and contact bifurcations) shows that the model is able to exhibit chaotic fluctuations. Moreover, long run equilibria without corruption cannot exist. Finally, corruption implies endogenous instability in the economic growth, due to periodic or aperiodic fluctuations. A further paper Brianzoni et al. (2010c) studies in depth the map from a mathematical point of view and describes the bifurcation curves of the superstable cycles.

The paper by Brianzoni, Coppier and Michetti, *Multiple equilibria in a discrete time growth model with corruption in public procurement*, makes the corruption level endogenous. More in detail, firms producing the public good differ with respect to their "reputation cost" deriving the fraction of firms which produce the low–quality public good by solving a one-shot game via the backward induction method. The resulting dynamic system describes the evolution of the capital per capita and of the corruption ratio. It admits multiple equilibria. The analysis of their stability and of the structure of their basins proves that stable equilibria with positive corruption may exist (according to empirical evidence), even though the State may reduce corruption by increasing the wage of the bureaucrat or by increasing the amount of tax revenues used to monitor corruption.

More recently, the paper by Brianzoni, Campisi and Russo, *Corruption and economic growth with non constant labour force growth*, extends the previous model by introducing endogenous labour force growth (described by the logistic equation). This leads to a three dimensional continuous piecewise system, where the state variables are the capital per capita, the corruption ratio and the population growth rate. The existence of an attractor (which may be strange) is proved. Moreover, numerical simulations are performed in order to obtain measures for the policy maker to fight corruption. To this regard, the role of several parameters is analyzed. A further research inherent to growth models with corruption, which is in progress, considers the evolution of non-compliant behaviour in public procurement.

2.1.3 Asset Pricing Models with Heterogeneous Expectations

Following a different line of research, asset pricing models with heterogeneous agents have been considered. As typical in the literature, different groups of agents have different expectations about future variables. For example, the paper by Brianzoni, Cerqueti and Michetti, *A dynamic stochastic model of asset pricing with heterogeneous beliefs*, studies an asset pricing model with heterogeneous expectations where the prevision rules depend on the proportion of agents belonging to the same group, weighted by a *confidence parameter*. The new ingredient has a stabilizing effect in the dynamic behaviour, since the system is globally stable if the confidence parameter is great enough. Differently, for small values of the confidence parameter, the system shows complex dynamics. In this last case, there exists a stability region which is analyzed both in the deterministic and in the stochastic framework.

The paper by Brianzoni, Mammana and Michetti, *Updating wealth in an asset pricing model with heterogeneous agents*, focuses on the wealth dynamics when two groups of agents (with different beliefs) populate the market. In fact, at all times, the wealth of each group is updated as a consequence of the switching mechanism. As a result, the final system is defined by a nonlinear, three-dimensional and continuous piecewise map, describing the evolution of the difference in the fractions of agents, the difference in the relative wealths and the fundamental price ratio. Although the complexity of the resulting system, the authors prove the existence of two types of steady states (fundamental and non fundamental equilibria) and perform the stability analysis of the fundamental steady state. Moreover, the paper shows how complexity is mainly due to the wealth dynamics. In the framework of heterogeneous agents, as a future development, we are interested in considering different ways to model heterogeneous expectations, taking into account empirical evidences that the related literature offers.

2.2   *Continuous Dynamical Systems with Delays*

The introduction of time delay in differential equations has been shown to be an efficient method for the modeling of nonlinear dynamics appearing in many complex phenomena of the applied sciences. What defines a time delay system is the feature that the system's future evolution depends not only on its present value but also on its past history. In recent times there has been a renewed interest in the development and analysis of delayed mathematical models in economics due to the role played by time delays in capturing more complex dynamics, thus enriching the description of the whole system. Time delays can be modeled in many different ways. The choice of the type has situation-dependency and implies the use of different analytical methods and techniques. There are two main types of delay: fixed time delay and continuously distributed time delay. The models that require fixed time delays make use of delay differential equations, whose characteristic equation is a mixed polynomial-exponential equation with infinitely many eigenvalues. The models with continuously distributed time delays have instead dynamic equations that are Volterra type integro-differential equations, where the characteristic equation is a polynomial equation with finitely many eigenvalues. In the following, some of our contributions on the subject are briefly reviewed (see Refs. Caraballo et al. 2018; Guerrini et al. 2018, 2019; Gori et al. 2018). For future research, we propose the introduction of stochastic terms in our models.

The paper by Caraballo, Colucci and Guerrini, *Dynamics of a continuous Hénon model*, deals with a continuous time version of the Hénon map, which is one of the most studied examples of discrete dynamical systems that exhibit chaotic behavior. In absence of time delays, the model's solutions either converge to the stable steady state or diverge. Hence, the complex dynamics of the discrete model is not displayed by the continuous one. On the other hand, if time delays are introduced in the system, then the complexity of the discrete model may be recovered. Through Hopf bifurcation and stability switches analysis, the authors show the appearance of stable limit cycles with increasing period and the presence of a strange attractor that resembles the famous Hénon attractor. The last part of the work concentrates on this attractor. In particular, the existence of a trapping region (positively invariant set) and of an absorbing set is proved. For future research, it is proposed the problem of multistability, i.e. the coexistence of several local attractors for the system.

The paper by Caraballo, Colucci and Guerrini, *On a predator prey model with nonlinear harvesting and distributed delay*, analyses a two predator prey model with nonlinear harvesting (Holling type II) with both constant and distributed delay. The aim is to study the impact of harvesting on a two species community in presence of delays. The authors consider the cases of a single delay and continuously distributed delays, respectively, and then show a variety of dynamics ranging from simple cyclic oscillations to complex behavior involving chaos. Following the tradition of the study on fixed delay equations, the authors start from the local stability analysis of the steady state and then consider the question of stability switching of the fixed delay. It is found that the model may become unstable with an increase of time delay, the number of stability switches is finite, and Hopf bifurcations may emerge. The authors then turn their attention to similarity and dissimilarity of dynamics generated under the fixed time delay to those generated under the continuously distributed time delay. It is found that if the delay kernel is a weak kernel the model exhibits an attracting fixed point or an attracting limit cycle (periodic orbit) generated by Hopf bifurcation, while if the delay kernel is a strong kernel then cycles with increasing periods occur.

The paper by Guerrini, Matsumoto and Szidarovszky, *Neoclassical growth model with multiple distributed delays*, extends the neoclassical model of Solow and Swan, which has been a prototype model for analyzing long-run economic growth, by assuming two independent and distributed delays: one in the nonlinear capital accumulation through savings and the other one in the capital depreciation. This modeling of time delays allows reduction of the dynamics to a set of ordinary differential equations. According to the Routh-Hurwitz stability criterion, it is found a condition under which a stationary state loses stability and bifurcates to a cyclic oscillation. In addition, by combining analytical methods and numerical experiments, the authors reveal the dual role of time delay in destabilizing or stabilizing the economy, depending on the combination of two delays. Such duality of time delay does not appear in one delay models.

The paper by Guerrini, Matsumoto and Szidarovszky, *Delay Cournot duopoly models revisited*, generalizes a duopoly model with best reply dynamics, special case of the Cournot oligopoly model proposed in Howroyd and Russel, where sufficient condition for stability are provided in cases each firm experiences delays in implementing information on its own output (implementation delay) and in collecting information on its competitors' outputs (information delay). Two Cournot duopoly with different delays are built in a continuous time framework. The first model (model I) includes equal implementation and information delays. The second model (model II) possesses instead only the information delays, i.e. the implementation delays are assumed to be zero. The stability of these models is investigated by modern analytical techniques, such as the method of stability switches curves, and by means of sophisticated numerical methods. It emerges that the delays have the dual roles of destabilizer and stabilizer in model I, while they do not affect stability in model II.

Therefore, these delay models may explain various dynamics ranging from simple to complex behavior under Cournot competition. The following model extensions are also proposed: the case of more than two firms in the model, the introduction in the duopoly case of more alternative specifications for the delays, the assumption of nonlinear price and/or cost functions in order to make the system nonlinear.

The paper by Gori, Guerrini and Sodini, *Time delays, population, and economic development*, proposes a growth model à la Solow augmented with time delays in technology and population dynamics, where population grow evolves according to a logistic-type equation with carrying capacity positively correlated with the accumulation of physical capital. The resulting model is a system of delay differential equations with at most three steady state equilibria. The stability and bifurcations of these equilibria are analyzed and the emergence of complex behaviors is demonstrated. More specifically, by making use of a mixture between analytical and numerical tools, the authors illustrate how the Solow model becomes able to explain the convergence towards a high equilibrium or a Malthusian trap as well as long-term fluctuations in income and population. This paper thus illustrates how the Solow model may offer insight into the understanding of some transmission mechanisms between economic and demographic variables, that are often difficult to identify in growth models belonging to the Unified Growth Theory due to the complexity of their structure.

References

Abínzano, I., Seco, L., Escobar, M., & Olivares, P. (2009). Single and double Black-Cox: Two approaches for modelling debt restructuring. *Economic Model, 26*(5), 910–917.

Alvarez, L. H. R., & Koskela, E. (2006). Irreversible investment under interest rate variability: Some generalizations. *Journal of Business, 79*(2), 623–644.

Anderson, R. W., & Sundaresan, S. (1996). Design and valuation of debt contracts. *The Review of Financial Studies, 9*(1), 37–68.

Andreoli, A., Ballestra, L. V., & Pacelli, G. (2015). Computing survival probabilities based on stochastic differential models. *Journal of Computational and Applied Mathematics, 277*, 127–137.

Andreoli, A., Ballestra, L. V., & Pacelli, G. (2016a). Pricing credit default swaps under multifactor reduced-form models: A differential quadrature approach. *Computational Economics, 51*, 379–406.

Andreoli, A., Ballestra, L. V., & Pacelli, G. (2016b). From insurance risk to credit portfolio management: a new approach to pricing CDOs. *Quantitative Finance, 16*, 1495–1510.

Baldi, F., & Trigeorgis, L. (2009). A real options approach to valuing brand leveraging options: How much is Starbucks brand equity worth? Working Paper, 2009.

Ballestra, L. V., & Pacelli, G. (2009). A numerical method to price defaultable bonds based on the Madan and Unal credit risk model. *Applied Mathematical Finance, 16*(1), 17–36.

Ballestra, L. V., & Pacelli, G. (2014). Valuing risky debt: A new model combining structural information with the reduced-form approach. *Insurance: Mathematics and Economics, 55*(1), 261–271.

Ballestra, L. V., Pacelli, G., & Radi, D. (2014). Valuing investment projects under interest rate risk: empirical evidence from European firms. *Applied Economics, 49*(56), 5662–5672.

Ballestra, L. V., Pacelli, G., & Radi, D. (2016). A very efficient approach for pricing barrier options on an underlying described by the mixed fractional brownian motion. *Chaos, Solitons & Fractals, 87*, 240–248.

Ballestra, L. V., Pacelli, G., & Radi, D. (2017). Computing the survival probability in the Madan-Unal credit risk model: Application to the CDS market. *Quantitative Finance, 17*(2), 299–313.

Bäuerle, N. (2002). Risk management in credit risk portfolios with correlated assets. *Insurance: Mathematics and economics, 30*(2), 187–198.

Bayraktar, E., & Young, V. R. (2007). Hedging life insurance with pure endowments. *Insurance: Mathematics and Economics, 40*, 435–444.

Bender, C., & Elliott, R. J. (2004). Arbitrage in a discrete version of the Wick-fractional Black-Scholes market. *Mathematics of Operations Research, 29*(4), 935–945.

Bernard, C., Le Courtois, O., & Quittard-Pinon, F. (2005). Market value of life insurance contracts under stochastic interest rates and default risk. *Insurance: Mathematics and Economics, 34*(3), 499–516.

Bernardo, A. E., Chowdhry, B., & Goyal, A. (2012). Assessing project risk. *Journal of Applied Corporate Finance, 24*(3), 94–100.

Black, F., & Cox, J. C. (1976). Valuing corporate securities: Some effects of bond indenture provisions. *The Journal of Finance, 31*(2), 351–367.

Brianzoni, S., Mammana, C., & Michetti, E. (2007). Complex dynamics in the neoclassical growth model with differential savings and non-constant labor force growth. *Studies in Nonlinear Dynamics and Econometrics, 11*(3), 1–17.

Brianzoni, S., Mammana, C., & Michetti, E. (2009). Non-linear dynamics in a business-cycle model with logistic population growth. *Chaos, Solitons & Fractals, 40*(2), 717–730.

Brianzoni, S., Cerqueti, R., & Michetti, E. (2010a). A dynamic stochastic model of asset pricing with heterogeneous beliefs. *Computational Economics, 35*(2), 165–188.

Brianzoni, S., Mammana, C. & Michetti, E. (2010b). Updating wealth in an asset pricing model with heterogeneous agents. Discrete Dynamics in Nature and Society, 676317.

Brianzoni, S., Michetti, E., & Sushko, I. (2010c). Border collision bifurcations of superstable cycles in a one-dimensional piecewise smooth map. *Mathematics and Computers in Simulation, 81*(1), 52–61.

Brianzoni, S., Coppier, R., & Michetti, E. (2011). Complex dynamics in a growth model with corruption in public procurement. *Discrete Dynamics in Nature and Society, 862396*, 1–27.

Brianzoni, S., Mammana, C., & Michetti, E. (2012). Variable elasticity of substitution in a discrete time Solow-Swan growth model with differential saving. *Chaos, Solitons & Fractals, 45*(1), 98–108.

Brianzoni, S., Coppier, R., & Michetti, E. (2015a). Multiple equilibria in a discrete time growth model with corruption in public procurement. *Quality & Quantity, 49*(6), 2387–2410.

Brianzoni, S., Mammana, C., & Michetti, E. (2015b). Local and global dynamics in a neoclassical growth model with non concave production function and non constant population growth rate. *SIAM Journal on Applied Mathematics, 75*(1), 61–74.

Brianzoni, S., Campisi, G., & Russo, A. (2018). Corruption and economic growth with non constant labour force growth. *Communications in Nonlinear Science, 58*, 202–219.

Briys, E., & de Varenne, F. (1997). Valuing risky fixed rate debt: An extension. *Journal of Financial and Quantitative Analysis, 32*(2), 239–248.

Caraballo, T., Colucci, R., & Guerrini, L. (2018). On a predator prey model with nonlinear harvesting and distributed delay. *Communications on Pure & Applied Analysis, 17*, 2703–2727.

Cathcart, L., & El-Jahel, L. (2003). Semi-analytical pricing of defaultable bonds in a signaling jump-default model. *Journal of Computational Finance, 6*(3), 91–108.

Cathcart, L., & El-Jahel, L. (2006). Pricing defaultable bonds: A middle-way approach between structural and reduced-form models. *Quantitative Finance, 6*(3), 243–253.

Charalampopoulos, G., Katsianis, D., & Varoutas, D. (2001). The option to expand to a next generation access network infrastructure and the role of regulation in a discrete time setting: A real options approach. *Telecommun Policy, 35*(9–10), 895–906.

Chen, C.-J., & Panjer, H. (2003). Unifying discrete structural models and reduced-form models in credit risk using a jump-diffusion process. *Insurance: Mathematics and Economics*, *33*(2), 357–380.

Cheridito, P. (2003). Arbitrage in fractional Brownian motion models. *Finance and Stochastics*, *7*(4), 533–553.

Crouhy, M., Galai, D., & Mark, R. (2000). A comparative analysis of current credit risk models. *Journal of Banking & Finance*, *24*(1), 59–117.

Dieci, R., He, X., & Hommes. C. (2014). *Nonlinear economic dynamics and financial modelling*. Springer International Publishing.

Ding, Z., Granger, C., & Engle, R. (1993). A long memory property of stock market returns and a new model. *Journal of Empirical Finance*, *1*(1), 83–106.

Dixit, A. K., & Pindyck, R. S. (1994). *Investment under uncertainty*. Princeton, NJ: Princeton University Press.

Duffee, G. R. (1999). Estimating the price of default risk. *The Review of Financial Studies*, *12*(1), 197–226.

Duffie, D., & Lando, D. (2001). Term structures of credit spreads with incomplete accounting information. *Econometrica*, *69*(3), 633–664.

Duffie, D., & Singleton, K. J. (1999). Modeling term structures of defaultable bonds. *The Review of Financial Studies*, *12*(4), 687–720.

Duncan, T. E., Hu, Y., & Pasik-Duncan, B. (2000). Stochastic calculus for fractional Brownian motion I Theory. *SIAM Journal on Control and Optimization*, *38*(2), 582–612.

Feng, R., & Volkmer, H. W. (2012). Modeling credit value adjustment with downgrade-triggered termination clause using a ruin theoretic approach. *Insurance: Mathematics and Economics*, *51*(2), 409–421.

Fernandes, B., Cunha, J., & Ferreira, P. (2011a). The use of real options approach in energy sector investments. *Renewable & Sustainable Energy Reviews*, *15*(9), 4491–4497.

Fernandes, B., Cunha, J., & Ferreira, P. (2011b). Real options theory in comparison to other project evaluation techniques. (Vol. 28–29).

Fontana, C., & Montes, J. M. A. (2014). A unified approach to pricing and risk management of equity and credit risk. *Journal of Computational and Applied Mathematics*, *259*, 350–361.

Franks, J. R., & Torous. W. (1989). An empirical investigation of U.S. firms in reorganization. *The Journal of Finance*, *44*(3), 747–769.

Giesecke, K. (2006). Default and information. *Journal of Economic Dynamics and Control*, *30*(11), 2281–2303.

Gori, L., Guerrini, L., & Sodini, M. (2018). Time delays, population, and economic development. *Chaos*, *28*, 055909.

Grundke, P., & Riedel, K. O. (2004). Pricing the risks of default: A note on Madan and Unal. *Review of Derivatives Research*, *7*(2), 169–173.

Gu, H., Liang, J.-R., & Zhang, Y.-X. (2012). Time-changed geometric fractional Brownian motion and option pricing with transaction costs. *Physica A: Statistical Mechanics and its Applications*, *391*(16), 3971–3977.

Guerrini, L., Matsumoto, A., & Szidarovszky, F. (2018). Delay cournot duopoly models revisited. *Chaos*, *28*, 093113.

Guerrini, L., Matsumoto, A., & Szidarovszky, F. (2019). Neoclassical growth model with multiple distributed delays. *Communications in Nonlinear Science*, *70*, 234–247.

Hao, R., Liu, Y., & Wang, S. (2014). Pricing credit default swap under fractional Vasicek interest rate model. *Journal of Mathematical Finance*, *4*(1), 10–20.

Hommes, C. H. (2013). *Behavioral rationality and heterogeneous expectations in complex economic systems*. Cambridge University Press.

Ingersoll, J. E, Jr., & Ross, S. A. (1992). Waiting to invest: Investment and uncertainty. *Journal of Business*, *65*(1), 1–29.

Jones, E., Mason, S., & Rosenfeld, E. (1984). Contingent claims analysis of corporate capital structures: An empirical investigation. *The journal of finance*, *39*(3), 611–625.

Leland, H. E., & Toft, K. B. (1996). Optimal capital structure, endogenous bankruptcy, and the term structure of credit spreads. *The Journal of Finance, 51*(3), 987–1019.

Liang, J.-R., Wang, J., Zhang, W.-Y., Qiu, W.-Y., & Ren, F.-Y. (2010). Option pricing of a bi-fractional Black-Merton-Scholes model with the Hurst exponent H in [1/2,1]. *Applied Mathematics Letters, 23*(8), 859–863.

Lim, S. C., & Muniandy, S. V. (2002). Self-similar Gaussian processes for modeling anomalous diffusion. *Physical Review E, 66*(2), 021114.

Lim, S. C., & Muniandy, S. V. (2003). Generalized Ornstein-Uhlenbeck processes and associated self-similar processes. *Journal of Physics A: Mathematical and General, 36*(14), 3961.

Lo, A. W. (1991). Long-term memory in stock market prices. *Econometrica, 59*(5), 1279–1313.

Longjin, L., Ren, F.-Y., & Qiu, W.-Y. (2010). The application of fractional derivatives in stochastic models driven by fractional Brownian motion. *Physica A: Statistical Mechanics and its Applications, 389*(21), 4809–4818.

Longstaff, F. A., & Schwartz, E. S. (1995). A simple approach to valuing risky fixed and floating rate debt. *The Journal of Finance, 50*(3), 789–819.

Loureiro, M. V., Claro, J., & Pereira, P. J. (2015). Capacity expansion in transmission networks using portfolios of real options. *International Journal of Electrical Power & Energy Systems, 64*, 439–446.

Madan, D. B., & Schoutens, W. (2008). Break on through to the single side. *The Journal of Credit Risk, 4*(3), 3–20.

Madan, D. B., & Unal, H. (1998). Pricing the risk of default. *Review of Derivatives Research, 2*(2), 121–160.

Madan, D. B., & Unal, H. (2000). A two-factor hazard rate model for pricing risky debt and the term structure of credit spreads. *Journal of Financial and Quantitative Analysis, 35*(1), 43–65.

Mandelbrot, B., & Van Ness, W. (1968). Fractional Brownian motions, fractional noises and applications. *SIAM Review, 10*(4), 422–437.

Manley, B., & Niquidet, K. (2010). What is the relevance of option pricing for forest valuation in New Zealand? *Forest Policy and Economics, 12*(4), 299–307.

McDonald, R., & Siegel, D. (1986). The value of waiting to invest. *The Quarterly Journal of Economics, 101*(4), 707–727.

Merton, R. C. (1974). On the pricing of corporate debt: The risk structure of interest rates. *The Journal of Finance, 29*(2), 449–470.

Myers, S. C., & Majd, S. (1990). Abandonment value and project life. *Advances in Futures and Options Research, 4*, 1–21.

Nadarajah, S., Margot, F., & Secomandi, N. (2015). Relaxations of approximate linear programs for the real option management of commodity storage. *Management Science, 61*(12),

Nadarajah, S., Secomandi, N., Sowers, G., & Wassick, J. (2017). Real option management of hydrocarbon cracking operations. In *Real options in energy and commodity markets*. Springer.

Øksendal, B. (2003). *Fractional Brownian motion in finance*. Dept. of Math: University of Oslo.

Rabinovitch, R. (1989). Pricing stock and bond options when the default-free rate is stochastic. *Journal of Financial and Quantitative Analysis, 24*(4), 447–457.

Rostek, S., & Schöbel, R. (2013). A note on the use of fractional Brownian motion for financial modeling. *Economic Modelling, 30*(1), 30–35.

Santos, L., Soares, I., Mendes, C., & Ferreira, P. (2014). Real options versus traditional methods to assess renewable energy projects. *Renewable Energy, 68*, 588–594.

Schoutens, W., & Cariboni, J. (2009). *Lévy Processes in Credit Risk*. Wiley.

Schulmerich, M. (2010). *Real options valuation: the importance of interest rate modelling in theory and practice*. Springer.

Vasicek, O. (1977). An equilibrum characterization of the term structure. *Journal of Financial Economics, 5*(2), 177–188.

Wang, A.-T., Ren, F.-Y., & Liang, X.-Q. (2003). A fractional version of the merton model. *Chaos, Solitons & Fractals, 15*(3), 455–463.

Wang, J., Liang, J.-R., Lv, L.-J., Qiu, W.-Y., & Ren, F.-Y. (2012). Continuous time Black-Scholes equation with trasaction costs in subdiffusive Brownian motion regime. *Physica A: Statistical Mechanics and its Applications, 391*(3), 750–759.

Wang, X.-T., Qiu, W.-Y., & Ren, F.-Y. (2001). Option pricing of fractional version of the Bblack-Scholes model with Hurst exponent H being in (1/3,1/2). *Chaos, Solitons & Fractals, 12*(3), 599–608.

Wang, X.-T., Liang, X.-Q., Ren, F.-Y., & Zhang, S.-Y. (2006). On some generalization of fractional Brownian motions. *Chaos, Solitons & Fractals, 28*(4), 949–957.

Young, V. R. (2008). Pricing life insurance under stochastic mortality via the instantaneous Sharpe ratio. *Insurance: Mathematics and Economics, 42*, 691–703.

Zhang, P., Sun, Q., & Xiao, W.-L. (2014). Parameter identification in mixed Brownian-fractional Brownian motions using Powell's optimization algorithm. *Economic Modelling, 40*, 314–319.

Zhou, C. (2001). The term structure of credit spreads with jump risk. *Journal of Banking & Finance, 25*(11), 2015–2040.

Productivity Differentiation and International Specialization of Firms and Districts

Marco Cucculelli, Alessia Lo Turco and Massimo Tamberi

Abstract This paper provides an overview of three lines of research that have been developed under an empirical micro-level perspective by scholars of the Ancona school of economics. First, the tendency toward an endogenous productivity dispersion in later developing economies is assessed from an entrepreneurial perspective, i.e. the amount and distribution of organizational and entrepreneurial abilities during the catching-up process. Second, the features and dynamics of a country specialization pattern are analyzed within the processes of upgrading of the production structures toward more complex and growth-generating models. Finally, the role and influence on firm performance of localization economies and different styles of corporate governance have been studied by framing them into the wider picture of firm organizational models and the hierarchical structure of market transactions. The paper concludes with a brief discussion of potential avenues for future research.

1 Introduction

Giorgio Fuà proposed an idea of economic development free from the stereotype of "development models" hitherto established in the empirical analysis. These models constituted a vulgarization of previous research themes, mainly originated from the work of Simon Kuznets, a pioneer of quantitative studies on economic growth and development.

M. Cucculelli · A. Lo Turco · M. Tamberi (✉)
Department of Economics and Social Sciences, Università Politecnica delle Marche, Piazzale Martelli 8, 60121 Ancona, Italy
e-mail: m.tamberi@univpm.it

M. Cucculelli
e-mail: m.cucculelli@univpm.it

A. Lo Turco
e-mail: a.loturco@univpm.it

© Springer Nature Switzerland AG 2019
S. Longhi et al. (eds.), *The First Outstanding 50 Years of "Università Politecnica delle Marche"*,
https://doi.org/10.1007/978-3-030-33879-4_10

The simplification of the original framework led some authors to reduce implicitly the development process to an almost deterministic phenomenon, in which the countries ended up following the path of development codified by their predecessors, despite having diversified historical roots and different periods of entry into the epoch of "modern economic growth" (Kuznets 1973).

Now, although the most advanced economies constitute a very powerful ideological and practical benchmark, the environment in which the economic takeoff and the subsequent phases of development happen is different by definition for a late developing country. While this is partly trivial, it actually hides a very relevant side: when compared to the areas of early development (basically, North America and Western Europe), late developing economies lag behind not only temporally, but, above all, in terms of the extant highly advanced technological-organizational frontier.

To put it in the words of Abramovitz (1986), this constitutes a "potential" for growth (the so-called backwardness advantages of Gerschenkronian memory), but, for its "realization", this potential must find a number of non-trivial conditions, including institutional ones (the so called "social capability", again in Abramovitz's words).

Moving along these lines, Fuà focused his interests on the peripheral areas of Europe and highlighted the endogenous character of the development process in general: the conditions and characteristics of economic development are modified by the process itself, and they are never equal to themselves over time.

It is important to stress that those considerations start from the implicit assumption that there is a substantial process of international integration, through which reciprocal conditioning takes place.[1]

The previous conceptual scheme has promoted a series of research lines that have been carried out by the economists of Ancona for a long time and until now. In particular, we emphasize three different, but still interconnected, strands of research:

(1)   the dualism (dispersion) of productivity and size of companies due to the limited amount and unequal distribution of organizational and entrepreneurial skills (the O-I factor, in Fuà's words);
(2)   the international specialization pattern;
(3)   the organizational model: firms and industrial districts.

In the following Sections, we will briefly illustrate the research contributions that have been developed over time along those lines by the researchers of Ancona. Despite the inclusion of several past works, this survey focuses specifically on the more recent papers.

---

[1] The gap in technology and productivity of the less developed economies could never be a "potential" in the absence of the possibility of international transmission of the technology itself.

2 Productivity Dualism

The idea was first presented in a study (Fuà 1980, 1978), published at the end of the seventies, on six lagged and peripherical western European countries, in which the author presented a limited but meaningful empirical evidence that in lagged economies there were larger productivity and wage gaps among macro-sectors and among manufacturing firms of different sizes; he then suggested a mechanism of "endogenous" differentiation of productivity, in "later developing countries", linked to the catching-up process: given the large international productivity gap and the relatively slowness of the catching-up process, as a consequence of a limited endowment of entrepreneurial and organizational capabilities, it is natural to find higher productivity gaps in later developing countries.

Among the scholars from Ancona, this idea was theoretically formalized by Crivellini (1983); subsequently, a detailed analysis for the Italian economy, distinguishing between three size classes of firms, was conducted by Conti and Modiano (2012): their results showed the particular weakness of Italian micro-enterprises, a less pronounced disadvantage for large companies, but also a substantial strength of medium-sized firms.

A confirmation of the "productivity dualism" hypothesis in later developing economies can be found in recent contributions, as in Syverson (2011, p. 237): while in USA "the plant at the 90th percentile of the productivity distribution makes almost twice as much output with the same measured inputs as the 10th percentile", we find "larger productivity differences in China and India, with average 90–10 TFP ratios over 5:1". Recently the previous findings have also been confirmed for a large set of countries, e.g. in Tamberi (2018). The reason for productivity dispersion should be found in the fact that forces, linked to the technological improvements, enhancing firm differences in term of productivity, act stronger in developing/catching-up countries (and this includes also factor intensities), while mechanisms reducing firm differences, particularly the selection process, could be weaker in those same countries.

Results, based on different functional forms, with different estimation strategies, and under several robustness checks, clearly support the idea that the productivity dispersion decreases with the level of development: the signs of the coefficient of per capita income, as determinant of productivity dispersion, are as expected and always significant.

Note that, the previous notes make reference to an economic environment characterized by a substantial international integration (globalization). In particular, there are many contributions showing that a strong shaping force of dispersion of productivity levels across firms is represented by international trade. The analysis of the evolution of firm productivity in this respect cannot avoid considering the increasing pace of international economic integration from the 90s onwards. The completion of the European Union, the EU Customs Union negotiation with the future Center and Eastern European member states, as well as an acceleration of regional trade integration in the Americas and Asia and the conclusion of the Uruguay Round with the institution of the WTO, all are events that have represented a progress towards increased global economic integration. The latter has been further strengthened by the entry of China in the WTO in December 2001. The concurrent increasing availability of firm level data has favored a detailed analysis of differences in productivity across firms differently involved in international markets and has led to a new theoretical paradigm explaining how more productive firms self-select into exporting and why trade liberalization in an industry favors disproportionately more firms at the top of the productivity distribution while favoring the exit of least productive firms out of the market (Melitz 2003; Bernard et al. 2003; Melitz and Ottaviano 2008). Trade liberalization would then represent a force towards higher average productivity levels and reduced productivity dispersion. In favor of this perspective is the study on Italy by Del Gatto et al. (2008) who show for the period 1983–1999 a reduction of productivity dispersion caused by trade openness.

Maggioni (2013) further shows for the 1998–2004 period that exposure to low income countries' final imports crowds out less efficient firms therefore reducing the heterogeneity in productivity in a sector, while imports from developed countries, which are associated to easier access for a wider variety of inputs, favor an increase in productivity dispersion. Indeed, beyond the composition effect stemming from exit of low productivity firms and the reshuffling of resources and sales in favor of high productivity firms, the literature has indicated some firm-specific effects which would stem from increased availability of greater and better variety of imports (Kashara and Rodrigue 2008). Indeed higher openness to imports of intermediates has been found for Italy to increase average productivity within industries (Lo Turco 2007).

However, when inspecting in greater detail the nexus between imports and productivity at the firm level over the period 2000–2004, Conti et al. (2014) find no role of firm intermediate imports on firm productivity. On the contrary, it seems that more productive firms self-select into importing and any positive productivity effects from importing fades away when controlling for firm export activity. This finding suggests that intermediate imports, more than affecting efficiency levels, help to maintain firms' competitive performance in the export market. The Italian specialization model focused on labor and manual intensive activities has been dramatically put under stress by the entry of low wage economies in the global economy.

The unprecedented boost in exports of low-income economies in the 2000s has mainly involved traditional goods, such as textiles, apparel, footwear and a wide array of further goods where manual intensive phases of production could be easily offshored in order to secure the maintenance of some specific and strategic upstream and downstream production phases of final goods at home in advanced economies. In this direction, Lo Turco and Maggioni (2013) find that only imports from cheap labor countries positively and significantly affect the export probability of Italian manufacturing firms. They interpret this finding as the working of the cost-saving channel, opposed to the technology channel, usually identified in the literature with imports from high-income countries. This evidence suggests that imports from low-income countries represent one of the key characteristics that allow firms to easily gain and preserve competitiveness in the export markets. If one also considers the recent evidence of learning by exporting originating from the analysis of Italian data (Serti and Tomasi 2008; Bratti and Felice 2012), a clear implication emerges: by means of cost saving imports positively affect the probability to export and this represents a source productivity gains originating from the firm's export activity.

This set of studies points in the direction of the weaknesses of the specialization model of the Italian economy in the framework of increased international competition from low income economies.

3 International Specialization

The Italian model of specialization is usually identified with "made-in-Italy", a mix of traditional consumer products and mechanics, in many cases of high quality. On this point two aspects should be emphasized.

The first concerns the positioning on traditional consumer goods: this is a consequence of the characteristics of being a "late comer" economy, which entails a certain comparative advantage in productions with relatively low labor costs (at least at certain stages of development); this also if the country was not and, above all, is not "absolutely" a late-comer; it is so if compared to the frontier countries, although it is itself advanced when compared to many backward world economies.

The second feature has to do with the strong integration between traditional and mechanical goods, especially machinery specialized for specific goods (i.e., machinery for furniture, clothing, etc.).

These aspects were investigated in Conti (1973), Alessandrini and Conti (1981), Conti and Menghinello (1998). Subsequently Tamberi (1999, 2007) showed how the evolution of the specialization model was linked, in an initial phase of growth, to the creation of the European common market, while the recent difficulties are a result of the expansion of competition to new emerging countries, especially in Asia. In any case, De Benedictis and Tamberi (2002) showed that the specialization structure of the Italian economy is highly differentiated, and that a clear evolution of the specialization was taking place since the end of the seventies. Nevertheless, we cannot expect a rapid "revolution" of the model of specialization, but, rather, a change along well defined directions, due to product and sector relatedness (Hausmann and Klinger 2007; Hausmann and Hidalgo 2011).

Moreover, we can stress that several scholars from Ancona contributed, especially in recent times, to new lines of investigation, that have become central in the international arena.

For example, on the basis on non-parametric estimation techniques, comparing different indices of overall specialization derived from the distribution of comparative advantages and controlling both countries' specificities and different nonparametric smoothing parameters, some papers (De Benedictis et al. 2009; Parteka and Tamberi 2013) discussed the subject of specialization evolution, and the results point out that countries diversify their product structure along their path of economic (structural) change. A summary of this literature, with several references to the contributions of Ancona researchers, can be found in Mau (2016).

Another modern and relevant field of analysis concerns the growing role of the service sector. Indeed, much of the importance of studying the model of specialization is based, today, on the increasing importance of services in world exports. In this respect, understanding the evolution of the manufacturing industries export performance is relevant to inferring what role Italy can play in the future of global trade, increasingly dominated by service producers. In this area, developed an empirical analyse to explore the determinants of the success of Italian business service firms in foreign markets. Beyond the role of experience at home and of their belonging to national and international networks, what emerged as a key factor was service firms' relationship with large industrial firms. In a further inspection of spillovers through vertical linkages in the service firms' internationalization process Conti, Lo Turco and Maggioni (2013) find that local export spillovers from downstream manufacturing buyers are important elements for service firms' export success.

The change of a country's specialization pattern is quite a complex phenomenon. The researchers from Ancona developed a number of analysis to study how countries at intermediate stages of development can cope with the need to upgrade their own production structures toward more complex and high growth generating productions. The empirical analyses adopted a micro-level perspective to inspect in detail how firms and locations move along the product space and what is the role of local and non-local capabilities in this process. The empirical setting of the manufacturing sector in Turkey in the 2000s has proved to be a particularly suitable setting of a country rapidly integrating into the global economy and experiencing several changes in the production structure. Lo Turco and Maggioni (2015a, 2016) first of all show that when starting to export, firms tend to expand their product scope by introducing more new products and there is a strong complementarity between starting to import and to export, as we find that purchasing inputs abroad reinforces the positive effects of export entry. Exploring the relative importance of firm- and local product-specific capabilities in fostering the introduction of new products, Lo Turco and Maggioni (2015a, 2016) find that firms' product space evolution is characterized by strong cognitive path dependence. From the spatial point of view, firm's internal specific resources turn to be more relevant for those firms operating in laggard regions, while firms in more developed regions can rely more on the availability of local technological-related competencies to enlarge their product space. However, in any case path dependence is strongly relaxed by some firm specific characteristics, such as firm size, efficiency and international exposure. In this respect, Lo Turco and Maggioni (2018) show that a relevant role in the accumulation of original product specific competencies is represented by foreign multinational affiliates active in the local market. They find that product discoveries—defined as products that are both new to the firm and to its location region—by firms are favored by their technological proximity to the product mix of co-located foreign firms. On the contrary product-specific competences spilling from co-located domestic firms and embedded in local imports do not play any role in the process of discovery. The main driver of these results seems to be the high intensity of local discoveries in novel and exclusive capabilities which foreign affiliates bring into the local economy. Further evidence in this direction is the work by Javorcik et al. (2015) which shows that Turkish firms in sectors and regions more likely to supply foreign affiliates tend to introduce more complex products, where complexity is measured according to Hausmann and Hidalgo (2009).

4 Entrepreneurship, Ownership and Industrial Districts

It is well known that the evolution of the Italian economy has been characterized by the presence of "Industrial Districts", a geographically settled collective process that incorporates both economic and social aspects. In this scenario, the unit of analysis has no longer been the individual or the company, but a nucleus of productive and social relationships rooted in the territory. In the past, the study of this phenomenon was largely discussed in relevant literature, partly from researchers of Ancona (Fuà and Zacchia 1983; Crivellini and Pettenati 1989).

This line of research was well embedded in the analysis of the features of the development process of late developing areas (in Italy the so-called NEC, North-East and Center regions). One of the limits for these countries to approaching the technological and organizational frontier was identified by Giorgio Fuà in the limited availability of the "O-I factor", i.e. the entrepreneurial (Imprenditoriale) and organizational (Organizzativo) capital. By stressing the importance of organizational and entrepreneurial skills, (i.e. O-I abilities), Fuà contributed to make entrepreneurship an integral part of the economic change and growth theory. Even though the centrality of the O-I abilities was fully recognized by classical economists, it has lately been largely neglected by mainstream economics because of the difficulty of framing it into neoclassic formal models. This trend has made the concept of entrepreneurial skills and ability almost disappear in the mainstream economic analysis: "the theoretical firm is entrepreneurless—the Prince of Denmark has been expunged from discussion of Hamlet" (Baumol 1968).

In late '70s, several contributions made the effort to model the entrepreneurial activity by stressing some of its basic constitutive features. Among these, the coordination of production and the size and scope of the business initiatives (Lucas 1978); the risk-taking behavior and the contractual structure of the firms (Kihlstrom and Laffont 1979); the ability to seize market opportunities and the innovation activity of the firm (Holmes and Schmitz 1990; Aghion and Howitt 1992). Despite these efforts, however, a model of individual real-world entrepreneur was not found in mainstream economics even in the following decades, as it eluded analytical tractability.

If the development of an analytical framework to model the entrepreneur was problematical, the empirical approach was more revealing. Following a well-rooted tradition of the applied studies in the economics of development, Fuà pointed to the availability of organizational and entrepreneurial abilities as a limit—or a driver—to the process of development and growth. Quoting Fuà (1980), "the several factors usually indicated to explain productivity and growth can be summarized into two main categories: organizational and entrepreneurial abilities on the one hand, and the productive capital on the other" (Fuà 1980, p. 29). Following this statement, a number of studies were conducted by the group of scholars of the Ancona school of economics, aimed at assessing the role of the entrepreneurial capital in support of the growth process of the firm and the economy. Without a sound analytical definition of the entrepreneur, the contributions from Ancona looked for proxies of entrepreneurial behavior of the company that could be elicited by characteristics of the firm or specific changes in its behavior.

Under the hypothesis that the owner/founder embodies the entrepreneurial attitude of the company, a first line of study was driven by the impact of the change of the firm ownership and management on performance. When the founder steps down, the firm experiences an organizational shock that forces the company to reshape and realign previous organizational routines into a new setting. This adjustment process—driven by an identifiable change in the stock of O-I factor—may be particularly intense and sometimes even lead the company to disappear if the adjustment is beyond the scope of the new owner/manager (Cucculelli and Micucci 2008). A second line of analysis used the equity ownership as a proxy for the entrepreneurial orientation of the company owners. Instead of considering a stylized firm where risk-averse managers make decisions in order to maximize the short period value of the company, as in mainstream theoretical economics, the empirical analysis has investigated the differences in the equity ownership of the company as a proxy for the company's behavior on the market, i.e. its entrepreneurial conduct (Cucculelli 2008). Among different types of ownership, the most diffuse and widespread in almost all countries and across all industries is family ownership. Because of their long-term orientation, family owners typically invest a significant amount of their wealth in the company and want to pass the firms down the generations. They value control over asset even more than performance, and this makes them highly sensitive to business risk conditions prevailing in the industry. More generally, their long-term orientation coupled with the concentration of family wealth and the desire to hand down the company through the generations make these owners very long-term oriented entrepreneurs. Conversely, as they are often scrutinized by the financial press and analysts on the ground of their short-term performance, financial owners are expected to have shorter investment horizons. This makes them less prone to invest into long term innovation programs and risk-taking activities, despite their excellent management practices and the ability to identify growth opportunities in the market.

The relevance of the company ownership and management as proxy for the O-I factors allows to extend the analysis to the debate on the industrial organization of production.

Since the early work on governance and development, scholars have commonly assumed that firms owned and controlled by a family are a source of competitive advantage in the early stages of industrialization, but also that it become a source of decline in the stages of mature development (Payne 1984; Ben-Porath 1980). The evidence on the presence of family firms around the world shows that this model of governance is actually among the most diffused worldwide.

A large literature shows that family firms are extremely well-placed to assist economic growth, as they combine several unique characteristics that make them extremely important across the entire lifecycle of the firm (Bertrand and Schoar 2006). Family ownership is a source of competitive advantage for economic development because of the support to the networks of small and medium-sized firms through family-specific advantages: long term orientation, greater access to internal financial capital, family social capital and stewardship. These features can be particularly relevant when the process of economic development is based on the predominance of micro- and small firms that build their competitiveness on a system of inter-firm relationships, as in industrial districts (IDs). In these spatial entities, concentrated populations of new and established family firms are embedded in social communities characterized by a 'homogenous system of values' (Becattini 1990). In IDs, family-based tacit knowledge and values can be created over long periods and transmitted into the wider community to facilitate low-cost coordination and regulate competition.

In this respect, a large body of literature has focused upon the importance of industrial districts to economic development. This issue was particularly relevant in the case of Italian post-war economic recovery (Amatori et al. 2013; Brusco and Paba 1997), as it was based on an exceptional proliferation of small, family-owned and managed firms. Recently, the challenges of globalization, accelerated innovation, the emergence of new technological paradigms and other major events have stimulated a number of structural modifications of the core structure of the system. Although the long-term effects of this restructuring phase are still hard to evaluate, one of the most important emerging phenomena is the increasing role of mid-size enterprises as they are demonstrating a strong ability to adapt and compete in the new scenarios. This would entail a shift towards a more hierarchical organization of the market transactions and a leading role of medium-sized firms, most of them still maintaining family ownership. In a series of papers, Tamberi (2001, 2007) evidenced a radical transformation of local systems, towards a more differentiated structure and, ultimately, with less accentuated "district" characteristics, as a consequence of the process of internationalization and of internal evolution of the systems themselves. As stressed also by several authors, the outcome has been that the "minimum unit of analysis" turned to firms in opposition of the previous "local system" dimension.

Research on medium-sized globally-oriented family firms suggests their relevance also in many other contexts (Storper 1993). Besides their role in Germany's postwar economic reconstruction (Herrigel 1996), Germany's *Mittelstand* firms are today one of the pillars of the present model of specialization of the country (Block and Spiegel 2013), with many 'hidden champions', highly-focused technology leaders that pursue global niche strategies (Simon 2009; Gedajlovic et al. 2012).

Some recent evidence from scholars of the school of economics of Ancona show the presence of a relevant direct impact of family ownership on the performance of Italian firms. When measured with financial metrics, family firms usually outperform other firms, with differences that emerge across the different size classes. Most significantly, the interplay between the "district effect" and the "family effect" changes significantly across the firm size distribution: while these two effects act as substitutes for smaller size firms, they are complements in medium-sized firms (Cucculelli and Storai, 2015). In particular, the inherent characteristics of family ownership in medium sized firms seem to provide the best mix to exploit the district advantage at best, by leveraging the benefits of the family governance and combining them optimally with the district organization of manufacturing activity.

This evidence connects to the observation of successful mid-size enterprises that can help developing a modern entrepreneurial culture (Fuà 1980; Coltorti 2009), in order to overcome the limits of traditional manufacturing localism. This happens through the adoption of new organizational models allowing them to take advantage of the SMEs excellent manufacturing capabilities, deeply-rooted in industrial districts, to serve a large and highly differentiated international demand. In detail, the competitive success of many medium size firms depends not only on the product—which remains a typical distinctive factor of *made in Italy*—but also on their capability to manage and control the entire value chain, from the research to the final market, and to leverage the resources and competences existing in the district. All these characteristics, that are typical of family ownership, play a crucial role in supporting strategies adopted by medium-sized firms: long-term orientation, reputation, commitment to local producers and to the local labor market, shared social capital, values and norms, and a general networking ability, which favors business alliances.

5   Conclusions: Reflections on Possible Future Research Trends

The digital revolution has led to two current and intertwined main vectors of change of the society and the economy: the increasing storage and availability of data and the rapid evolution of Artificial Intelligence (AI). There is a lively and growing debate on those issues. It is then natural to try to foresee how these features of the current economic era will affect the evolution of research on the three main fields discussed above.

In the past, Morgan (1988) observed that half of the articles published in the *American Economic Review* and the *Economic Journal* contained no data, while in physics journals the number was just 12% and in chemistry approximately zero. Instead, in recent times "almost half of these economists focus on empirical research; one third do a mixture of theory and applied work; only one fifth do entirely, or almost entirely, purely theoretical research" (Oswald and Ralsmark 2008).

Nevertheless, as stressed by Krugman (2018), "data almost never speak for themselves". In other sciences, with a more stable theoretical corpus, as in physics, this could be less important; economics, despite many efforts, still has an evolving interpretive apparatus, also because the theoretical "laws" are historically rooted and conditioned by many changes in the environment that they should explain. It may be possible that the overuse of data, sooner or later, will tend to re-evaluate the theory; this trend could also be coupled with "more eclecticism, more economic history" and "more emphasis on behavioral economics" (Baumol 1991).

In any case, the larger and growing data availability will confirm in the future the tendency of an increasing presence of empirical analysis. In particular, an important role will be played by the ever-growing availability of microdata.

In this respect, the investigation on the nexus between productivity diversification and development will take advantages of the growing availability of microdata at international level; we expect a larger possibility of measuring productivity diversification starting from firm and/or plant data, while the level of development will be referred at sub-national level. This will reinforce the possibility of finding clear results, also disentangling different effects and shedding light on the current and strong forces of spatial divergence within nations. The increasing availability of transaction level data will further develop the understanding of the micro level forces behind countries' specialization patterns as well as the availability of data on the local network of firms will help the understanding of the mechanics of the heterogeneous resilience of industrial and service districts in the context of an increased international competition.

This successful competition, for local economies, has to do with the problem of how to make them "sticky" locations in a rapidly changing space (Markusen 1996). This difficulty is also connected with the necessity of changing the product/qualitative specialization of plants/firms/local systems.

A first point is that the rapid spread of advanced technologies, linked to robotics and AI in both services and manufacturing production, further poses the question on the possible future specialization patterns of both developed and developing nations (and locations in general). There is a growing debate on the effects of advanced technologies on both high and low skilled labor tasks and, as such, this widens the competitive pressure on the labor markets of both developed and developing economies. The spread of new technologies boosts unprecedent firm level gains in productivity and stresses the importance of individual talent and creativity in high managerial positions. Moreover, this rapid spread may constitute a complement of existing global value chains and favors the rapid spatial concentration of manufacturing production.

For the specialization issue, the intriguing topics then will concern research on the factors stimulating the progressive change and upgrading of the model. If new technologies can perform both high and low skill intensive tasks, it will be of relevance the identification of which are the most effective investments in human capital for a country. In general, it is possible that investigations on productivity and specialization will be framed in a perspective in which the interaction between the positioning both in the "product space" and in the "geographical space" will be important. This interaction constitutes a key determinant for the formation of internal and external (local) competencies (O-I factor), and it will influence both the firm and local ("district") possibility of successfully competing in the international arena.

Evidently, local economies are involved in the Global Value Chains and their success will be linked to the capacity of their actors to be central players in the global network, and to connect with other central players. Nonetheless, capital is footloose and rapidly moves across space. The challenge for research is to highlight what are the main factors behind the uniqueness of specific production districts and to shed light on the elements favoring the conversion and resilience of existing districts in the context of a rapidly changing global economy.

In the case of Italy, also considering the atypical specialization of the country, the previous considerations are particularly relevant, since the Italian specialization problem is also linked to the "industrial district" type of industrial organization and to the presence a much-diversified set of small and medium-size emerging and/or successful firms.

We believe that, conditional on data availability, it will be important to understand and analyze the specific channels, the policy options and the opportunities for moving local economies and companies toward a more central position in the GVCs.

References

Abramovitz, M. (1986). Catching up, forging ahead, and falling behind. *The Journal of Economic History, 46*(2), 385–406.

Aghion, P., & Howitt, P. (1992). A model of growth through creative destruction. *Econometrica, 60*(2), 323–351.

Alessandrini, P., & Conti, G. (Eds.). (1981). *Commercio estero ed allargamento della CEE: Prospettive per l'industria italiana*. Il Mulino.

Amatori, F., Bugamelli, M., & Colli, A. (2013). Italian firms in history: size, technology and entrepreneurship. In G. Toniolo (Ed.), *The Oxford handbook of the italian economy since unification*. New York: Oxford University Press.

Baumol, W. J. (1968). Entrepreneurship in economic theory. *American Economic Review, 58*(2), 64–71.

Baumol, W. J. (1991). Toward a newer economics: The future lies ahead! *The Economic Journal, 101*(404), 1–8.

Becattini, G. (1990). The Marshallian indutrial district as a socio-economic notion. In F. Pyke, G. Becattini, & W. Sengenberger (Eds.), *Industrial districts and inter-firm co-operation in Italy*. Geneva: International Institute for Labour Studies.

Ben-Porath, Y. (1980). The F-connection: Families, friends and firms and the organization of exchange. *Population and Development Review, 6*(1), 1–30.

Bernard, A. B., Eaton, J., Jensen, J. B., & Kortum, S. (2003). Plants and productivity in international trade. *American Economic Review, 93*, 1268–1291.

Bertrand, M., & Schoar, A. (2006). The role of family in family firms. *Journal of Economic Perspectives, 20*(2), 73–96.

Block, J., & Spiegel, F. (2013). Family firm density and regional innovation output: An exploratory analysis. *Journal of Family Business Strategy, 4*(4), 270–280.

Bratti, M., & Felice, G. (2012). Are exporters more likely to introduce product innovations? *The World Economy, 35*(11), 1559–1598.

Brusco, S., & Paba, S. (1997). Per una storia dei distretti industriali italiani dal secondo dopoguerra agli anni novanta. In F. Barca (Ed.), *Storia del capitalismo italiano dal dopoguerra ad oggi*. Roma: Donzelli Editore.

Coltorti, F. (2009). Medium-sized firms, groups and industrial districts. An Italian perspective. In G. Becattini, M. Bellandi, & L. De Propris (Eds.), *A handbook of industrial districts*. Cheltenham: Edward Elgar.

Conti, G. (1973). Progresso tecnico e competitività internazionale nella esperienza italiana. *Moneta e Credito, 26*(104), 336–361.

Conti, G., Lo Turco, A., & Maggioni, D. (2013). Rethinking the import-productivity nexus for Italian manufacturing. *Empirica, 41*(4), 589–617.

Conti, G., & Menghinello, S. (1998). Modelli di impresa e di industria nei contesti di competizione globale: L'internazionalizzazione produttiva dei sistemi locali del made in Italy. *L'Industria, XIX*(2), 315–348.

Conti, G., & Modiano, P. (2012). Problemi dei paesi a sviluppo tardivo in Europa: Riflessioni sul caso italiano. *L'industria, XXXIII*(2), 221–235.

Crivellini, M. (1983). Vincoli organizzativi-imprenditoriali allo sviluppo: Una stilizzazione dell'approccio di Ancona. *Rassegna Economica, XLVII*(2), 361–412.

Crivellini, M., & Pettenati, P. (1989). Modelli locali di sviluppo. In G. Becattini (Ed.), *Modelli locali di sviluppo*. Il Mulino.

Cucculelli, M. (2008). Owner identity, corporate performance and firm growth. *Evidence from European Companies, Rivista di Politica Economica, 98*(2), 149–178.

Cucculelli, M., & Micucci, G. (2008). Family succession and firm performance: Evidence from Italian family firms. *Journal of Corporate Finance, 14*(1), 17–31.

Cucculelli, M., & Storai, D. (2015). Family firms and industrial districts. *Journal of Family Business Strategy, 6*(4), 234–246.

De Benedictis, L., & Tamberi, M. (2002). Il modello di specializzazione italiano: normalità e asimmetria. In M. Cucculelli & R. Mazzoni (Eds.), *Risorse e competitività*. Franco Angeli.

De Benedictis, L., Gallegati, M., & Tamberi, M. (2009). Overall specialization and income: Countries diversify. *The Review of World Economics—Weltwirtschaftliches Archiv, 145*(1), 37–55.

Del Gatto, M., Ottaviano, G. I., & Pagnini, M. (2008). Openness to trade and industry cost dispersion: Evidence from a panel of Italian firms. *J Regional Sci, 48*(1), 97–129.

Fuà, G. (1978). Lagged development and economic dualism. *Banca Nazionale del Lavoro Quarterly Review, XXXI*(125), 123–134.

Fuà, G. (1980). *Problems of lagged development in OECD Europe: A study of six countries*. Document 2277, OECD.

Fuà, G., & Zacchia, C. (1983). *Industrializzazione senza fratture*. Bologna: Il Mulino.

Gedajlovic, E., Carney, M., Chrisman, J. J., & Kellermanns, F. W. (2012). The adolescence of family firm research: taking stock and planning for the future. *Journal of Management, 38*(4), 1010–1037.

Hausmann, R., & Hidalgo, C. (2011). *The ATLAS of economic complexity—mapping paths to prosperity*. Cambridge, MA: Harvard-CID and MIT-Media Lab.

Hausmann, R., & Klinger, B. (2007). *The structure of the product space and the evolution of comparative advantage*. Harvard University CID working paper 146.

Herrigel, G. (1996). *Industrial constructions: The sources of German industrial power*. Cambridge: Cambridge University Press.

Holmes, T., & Schmitz, J. (1990). A theory of entrepreneurship and its application to the study of business transfers. *Journal of Political Economy, 98*(2), 265–294.

Javorcik, B., Lo Turco, A., & Maggioni, D. (2015). New and improved: Does FDI boost production complexity in host countries? *Economic Journal, 128*(614), 2507–2537.

Kasahara, H., & Rodrigue, J. (2008). Does the use of imported intermediates increase productivity? Plant level evidence. *Journal of Development Economics, 87*(1), 106–118.

Kihlstrom, R., & Laffont, J. (1979). A general equilibrium entrepreneurial theory of firm formation based on risk aversion. *Journal of Political Economy, 87*(4), 719–748.

Krugman, P. (2018). What do we actually know about the economy? (Wonkish), The New York Times, Sept. 16

Kuznets, S. (1973). Modern economic growth: findings and reflections. *American Economic Review, 63*(3), 247–258.

Lo Turco, A. (2007). International outsourcing and productivity in Italian manufacturing sectors. *Rivista Italiana degli Economisti, 12*(1), 125–146.

Lo Turco, A., & Maggioni, D. (2013). On the role of imports in enhancing manufacturing exports. *The World Economy, 36*(1), 93–120.

Lo Turco, A., & Maggioni, D. (2018). Local discoveries and technological relatedness: The role of MNEs, imports and domestic capabilities. *Journal of Economic Geography*, https://doi.org/10.1093/jeg/lbv024 (forthcoming).

Lo Turco, A., & Maggioni, D. (2016). On firms' product space evolution: the role of firm and local product relatedness. *Journal of Economic Geography, 16*(5), 975–1006.

Lo Turco, A., & Maggioni, D. (2015a). Imports, exports and the firm product scope: Evidence from Turkey. *The World Economy, 38*(6), 985–1015.

Lucas, R. (1978). On the size distribution of business firms. *Bell Journal of Economics, 9*(2), 508–523.

Maggioni, D. (2013). Productivity dispersion and its determinants: The role of import penetration. *Journal of Industry, Competition and Trade, 13*(4), 537–561.

Markusen, A. (1996). Sticky places in slippery space: A typology of industrial districts. *Economic Geography, 72*(3), 293–313.

Mau, K. (2016). Export diversification and income differences reconsidered: The extensive product margin in theory and application. *The Review of World Economy, 152*(2), 351–381.

Melitz, M. J. (2003). The impact of trade on intra-industry reallocations and aggregate industry productivity. *Econometrica, 71*(6), 1695–1725.

Melitz, M. J., & Ottaviano, G. I. P. (2008). Market size, trade, and productivity. *Review of Economic Studies, 75*(1), 295–316.

Morgan, T. (2008). Theory versus empiricism in academic economics: update and comparisons. *Journal of Economic Perspectives, 2*(4), 159–164.

Oswald, A. J., & Ralsmark, H. (2008). *Some evidence on the future of economics*. Economic research papers 269790. University of Warwick, Department of Economics.

Parteka, A., & Tamberi, M. (2013). What determines export diversification in the development process? Empirical assessment. *The World Economy, 36*(6), 807–826.

Payne, P. L. (1984). Family business in Britain: An historical and analytical survey. In A. Okochi & S. Yasuoka (Eds.), *Family business in the era of industrial growth: Its ownership and management*. Tokyo: University of Tokyo Press.

Serti, F., & Tomasi, C. (2008). Self selection and post-entry effects of exports: Evidence from Italian manufacturing firms'. *Review of World Economics, 144*(4), 660–694.

Simon, H. (2009). *Hidden champions of the 21st century: Success strategies of unknown world market leaders*. London: Springer.

Storper, M. (1993). Regional 'worlds' of production: learning and innovation in the technology districts of France, Italy and the USA. *Regional Studies, 27*(5), 433–455.

Syverson, C. (2011). What determines productivity? *Journal of Economic Literature, 49*(2):326–365.

Tamberi, M. (1999). Competitività, crescita e localizzazione in un settore tradizionale. *l'industria, 1*, 93–126.

Tamberi, M. (2001). Trasformazioni produttive nei sistemi locali della Marche. In G. Becattini, M. Bellandi, G. Dei Ottati, F. Sforzi (Eds.), *Il caleidoscopio dello sviluppo locale*. Firenze: Rosenberg & Sellier.

Tamberi, M. (2007). Crescita, competizione internazionale e trasformazioni strutturali nei sistemi locali In M. Moroni (Ed.), *Lo sviluppo locale*. Storia, economia, sociologia, il Mulino.

Tamberi, M, (2018), *Productivity differentiation along the development process: A "meso" approach* (p. 427). Quaderni di ricerca, Dipartimento di Scienze Economiche e Sociali, Università Politecnica delle Marche.

Sustainability of Real and Financial Markets

A. Manelli, S. Branciari, L. Montanini and A. D'Andrea

Abstract Over the past decade, agricultural commodity prices underwent wild swings. Periods of high growth, 2008, were followed by periods of sudden decrease, 2009. The excessive increase in food price volatility is a subject of great interest because it has to do with the survival of mankind. Recent academic studies and policy makers have reached sharply different conclusions about the dynamics of such fluctuations. Also, some of them have indicated the main cause in speculative transactions and in the little regulation of futures markets. Then, this paper analyzes whether agricultural prices increases are due to increased speculative activity, otherwise to factors relating specifically to the commodities markets. The results of this preliminary analysis do not show a direct relationship between the increase in speculative transactions and spot prices, rather a more developed future market can be of benefit also to spot prices, as argued in many academic studies. Further statistical analyses tend to corroborate these conclusions. In fact, the first results of cointegration analysis show that there is no evidence that financial derivative instruments determine the fluctuation of wheat future price.

1 Introduction

Agriculture is the basis of human life. The serious economic crisis—which struck many countries worldwide from the Lehman Brothers bankruptcy—showed a rise in prices of agricultural products. In one year—between 2010 and 2011—in the US

A. Manelli (✉) · S. Branciari · L. Montanini · A. D'Andrea
Department of Management, Università Politecnica delle Marche, Ancona, Italy
e-mail: a.manelli@univpm.it

S. Branciari
e-mail: s.branciari@univpm.it

L. Montanini
e-mail: l.montanini@univpm.it

A. D'Andrea
e-mail: a.dandrea@univpm.it

© Springer Nature Switzerland AG 2019
S. Longhi et al. (eds.), *The First Outstanding 50 Years
of "Università Politecnica delle Marche"*,
https://doi.org/10.1007/978-3-030-33879-4_11

market the Wheattkr Spot Index grew by 56.23%, while Gxgrwpsp Forward Index rose by 44.51%. During the same period, the FAO Food Price Index rose with an increase of 38% and exceeded the maximum recorded in 2008. In the same year, the price of major agricultural commodities such as wheat and corn, rose to a level almost four times higher than in 2007 and in 2009 it is suddenly halved. Price fluctuations can be considered intrinsic in the commodities market because these are linked to the real availability of agricultural raw materials, i.e. the size of supply and demand, as well the variables that affect them (weather events, climate change and energy prices). However, the excessive increase in price volatility—both rising and decreasing, with sudden fluctuation, even within the same stock market session—has contributed to increasing instability and uncertainty in the agricultural commodity market. The years 2006–2008 and 2010 were periods in which prices rose to worrying levels, especially for those poorest countries that spend most of their incomes on food.

Agricultural failures and consequent policy responses, also, can lead to big swings in prices and in dangerous political situations. At the same time, resource contention, including food, can result in tensions between Nations, since the food chains are becoming increasingly international and countries are increasingly dependent on each other. Accordingly, it follows that the subject of the investigation is the determinants of agricultural prices. This article examines what happens to the prices of real goods—specifically grain—if we assume a different commercial or speculative use of derivatives made from the grain itself as underlying. This study investigates the link between spot and futures prices, weather financial markets influence the real markets, as stated by journalists and politicians, and, if it is so, what is the magnitude of such links and how it is achieved.

For this purpose, later, in the article we analyze the concepts of volatility and speculation from an academic perspective, then, it analyzes more accurately the price trends and the factors that influence them.

2 Volatility, Speculation and Index Funds

Volatility is the change of price in a certain time interval, i.e. the price dispersion around their average. In periods of high volatility, prices experienced a pronounced swing. However, in low volatility periods, prices change more moderately. Usually, in times of high volatility—like in 2008—harm both producers and consumers. On the side of the latter, if the price transmission from agricultural products to food tends to be moderate for developed countries, this cannot be said for developing countries, which becomes a factor that causes significant problems; on the side of production, volatility causes greater risk because farmers postpone investments as they are unable to foresee future revenues.

Volatility is an indicator to assess the market's variability and uncertainty. Not all changes are a problem, as when prices show a uniform and well-established trend, or when they follow a seasonal pattern. The changes become problematic when they are large, unpredictable and unexpected, because they create a level of

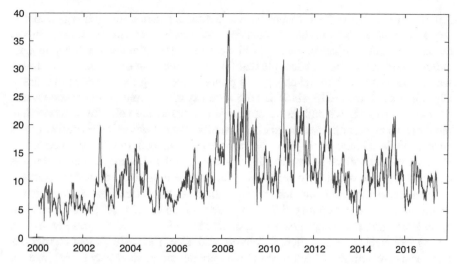

Fig. 1 Grain volatility. *Source* igc.com

uncertainty that increases the risk. Also, price changes related to external factors of the specific commodity, can be problematic as it is difficult to interpret. Using a historical analysis, it is easy to see how price variability tends to remain fairly stable over the long-run, while it is subject to rapid increases, in shorter periods.

Figure 1 shows the daily volatility of the IGC GOI (HV20) index, calculated as the standard deviation expressed in percentage terms.

Often, these are short-term increments, a few months, but sometimes it can last longer. This has happened since 2007. Moreover, commodities prices are interdependent with some mutual correlation due to real or financial ties. For example, the oil price increase produces the growth of transport costs, which if they are of considerable magnitude, are transferred to the final price of grain and its derivatives. This interdependence is called 'price transmission' because the shock found in a product is transmitted to others. So, even if there isn't a real or financial link between two assets one can still see common price dynamics. The common interdependence factors can be both observable variables—such as interest rates— and latent factors. These latter can be the expression of market fundamentals, such as a generalized growth in long-term demand, which isn't offset by the production growth (Gilbert 1995).

In this context, many have founded in the speculation[1] the root causes of these increases. Singleton (2014) sees speculative activity as the determinant of fluctuation in commodity prices. Analyzing the way that investors can change the price and focusing mainly on future prices Brunetti et al. (2015), show the existence of a link—albeit minor—between speculators and volatility, however, emphasizing how investment funds facilitate pricing and reduce market instability. Gilbert

---

[1]In the financial sector, in the common language, speculation suggest "a commercial transaction that seeks to earn a profit on the difference between current and future prices".

(2010) stresses that index investments was the main channel through which monetary and financial activities have influenced food prices in recent years. He points to the increase in the prices of energy products and metals stimulated interest in agricultural derivatives and he conclude that this diversification of financial investment was so wide as to determine changes in prices. Analyzing the agricultural market, Chevallier and Ielpo (2013) show that the volatility in the commodities market is less than other asset classes and, in the agricultural market, the volatility is lower than the precious metal or petroleum market. Usually, when speculation on derivatives is accused to destabilize prices we refer to spot prices: indeed, the futures speculation changes the derivatives prices that, in turn, changes the spot market. However, the changes may also result from variation in levels of storage and production. Accordingly, most academics do not fund evidence about possible alteration that speculation would produce on the commodities post market. Indeed, Kim (2015) suggests that speculators have had a significant and positive influence on the commodity market, during its recent period of financialization, and, also, she believes that restrictions on speculative activity in the futures market are not an efficient way to stabilize the commodity market. In her work, using a cross sectional analysis method, she shows that the relationship between price growth and the number of speculative positions is absent or negative and she finalizes that speculators reduce price growth rather than tighten them up. With regards to the agricultural market, she shows how the hedgers' activity have had a greater effect on the prices, rather than the speculators activity; the speculators presence shortens volatility and prevents high real market swings, as well as producing efficiency and liquidity. Ott (2014) tracks down the main determinant of high volatility in the relationship between closing stock and consumption of food goods—investigating the causes of agricultural price volatility. In this sense, he points out that the speculative activity and the liquidity have had a dampening effect on spot prices in the agricultural derivatives market, whereas, he believes that oil prices and exchange rates are factors that affect volatility. The concomitant phenomenon of strong price fluctuations and of growth in index fund[2] activity required an explanation. According to many observers, the fact that in the agricultural markets the increase in the number of transactions coincided with the sudden rise in prices prove that these transactions push agricultural prices to very high levels and to influence strongly the volatility—i.e. price fluctuations. Consequently,

---

[2]The index funds are funds that replicate the movements of specific market indices. The objective of this financial instruments is triple: the purpose of these financial instruments is to optimize the risk-performance ratio, to maintain a low correlation with traditional investment and to obtain positive returns, regardless of market trends through activities unrelated to original investment. In the raw materials field, these funds hold a certain number of differently weighted raw materials. During the purchase or sell of units, it is important to best represent the index. Generally, fundamentals are ignored. Index funds can exert pressure on agricultural market prices in different ways, often, although, agricultural commodities do not constitute most of these tools. In fact, when raising the price of non-agricultural raw materials, for example oil, the managers need to purchase agricultural products so that the balance between the different units is fair again. This strengthens the link between the different investment categories. In summary, a commodity index fund is a financial asset that invests on different futures or swaps markets with the aim of replicating profits in a price index.

some seek greater regulation of agricultural markets and of electronic markets, as well as the prohibition to make speculative transactions of considerable size. Is there a correlation? Although many observers agree on the causes of the price evolution, the empirical evidence is incomplete and not very conclusive. There is no doubt that in recent years the financial markets activity has increased in agriculture. However, to conclude that the rise in prices and their volatility are a direct consequence of this evolution is premature. Bauer and Minsch (2013) still show as tangible tests of the tie between speculation and markets volatilities have not emerged, rather they reiterate that financial market speculation has improved the agricultural market functioning. Critics describe the investment in index funds as excessive speculation. This is because long-term fund strategies may affect the balance between supply and demand pushing prices upwards. Masters and White (2008) show how speculators, such as index funds, following rising price performance, even taking a long position, threaten to intensify further growth. In 2008, according to the Commodity Future Trading Commission the share of index funds on commodity index investments was 24%. The huge and rapid increase of investments in commodity index funds seem to be related to the rising prices in the commodity markets. This correlation is known as 'Masters' hypothesis'[3] (Master 2009): he said, earlier, before the US Congress, and then, in front of the CFTC that speculations on commodity indices strongly influence future prices and distance them from fundamentals. This observation, which is not statistically proven, helped to reinforce the belief that passive speculators have caused a bubble not based on real economic magnitudes in the period of commodity prices volatility.

But assuming an efficient market, the index funds operating in the agricultural market produce a price stabilizing effect, as a direct consequence of their investment strategy. Index Funds, supporting future prices convergence towards spot prices expected, allow the decrease in the risk premium. As Prehn et al. (2014), even if the Fund pushes the future prices beyond the expected spot prices, the losses will bring some funds out of the market, correcting the imbalance. Particularly, the authors argue that hedge funds—having a long-term horizon—stabilize prices both by selling contracts that have increased their value and buying ones that have decreased and

---

[3]M. W. Masters, hedge funds manager and Masters Capital Management LLC founder, in May 2008, suggests that the determinant of increases—in 5 years—is to be identified in Corporate and Government Pension Funds and in other institutional investors, that he defines as a whole 'Index Speculator'. These speculators 'buy futures and then roll their position by buying calendar spreads. They never sell'. In this way, they consume liquidity and do not provide benefits to the futures markets. Regarding the correlation between their agricultural derivatives market entrance and periods of sudden price increases, Masters specifies that, after the severe equity bear market of 2002, this investors category began to look to the commodity futures market, pouring a liquidity on the main agricultural futures indices, speculating upward. Their strategy has meant that 'assets allocated to commodity index trading strategies have risen from $13 billion at the end of 2003 to $260 billion as of March 2008', and the prices of the index commodities 'have risen by an average of 183% in those five years'. In Masters opinion, commodity futures prices are the benchmark for the prices of actual physical commodities, so when Index Speculators, with their actions, 'drive futures prices higher, the effects are felt immediately in spot prices and in the real economy'. Therefore, Masters is asking US Congressional intervention to restrict these operations.

boosting competitive pressure on the same speculators. As the intensified funds work increases liquidity and decreases risk premium, fund activity allows farmers and commercial traders to hedge against price risks on more favorable terms. Moreover, the risk premium reduction, pushing farmers to store part of their crops, mitigates price fluctuations with a positive effect on spot markets too. Gilbert (2010) demonstrates that the common demand shocks are the factors that better explain the widespread price movements; he suggests index investments may be the channel through which these factors affect futures prices.[4] However, the effects of index investment for the hedgers, but also for consumers, are positive owing to a lower risk premium and more stable prices due to increased liquidity in the future markets.

2.1 Futures Market

The phases of price volatility might have affected the relationship between spot and future prices. For example, Baldi et al. (2013) show that, during periods when the bubbles are formed, the spot market becomes more important for pricing of certain agricultural commodities. This is due to the assumption that, during speculative phases, commodity prices are less related to financial trading and more dependent on fundamental models. Ivanov (2011) and Adammer et al. (2016) show that the pricing of future markets depends on their liquidity. If the volume traded on the futures market is low, then its contribution is low. In addition to liquidity, the degree of influence they exert on futures pricing could be determined by market turmoil. Considering this, the conclusion of Irwin et al. (2010) seems acceptable. To them, the Masters hypothesis is only superficially convincing due to defects and inconsistencies with the operational mechanisms of the futures markets. Indeed, they say that investing in futures markets may not be as demanding to the physical quantity of goods, i.e. the financial flow may not influence the price of raw materials for which futures were created. Indeed, a large part of a futures contract is creating a certain fixed price level.

Following this, commodity prices may change only if they change fundamental information perceived by investors. The simultaneous variation in spot and futures prices may indicate a correlation, but it is much harder to establish a cause-effect relationship.

2.2 Futures and Index Funds

Another way in which derivatives markets affect spot prices is through the bond index funds that have futures prices—it is considered an approximation in spot prices. However, to influence commodity prices, the funds should handle physical quantity of goods directly, since their market clearing price is only identified in spot markets

---

[4]E.g. when the funds are used as a hedge in dollar fluctuations.

where physical goods are bought and sold. Only their supply and demand become relevant to the market price. Indeed, Index Funds invest in futures market trading financial instruments and no physical products. This seems to deny the bubble theory in commodity markets, but some have found some statistical evidence about their effect.

Hamilton and Wu (2015), applying an OLS[5] regression model, from 2006 to 2012, for 12 agricultural commodities, indicate the absence of index funds on the agricultural products futures. To further confirm, the data provided by the CFTC reports show how some operators, in the futures markets, are 'pension funds or other managed funds taking a direct position in the futures contracts', while 'the majority represent positions by swap dealers, who offer their clients an over-the-counter product that mimics some futures-based index'. The clearest evidence is provided by Tang and Xiong (2012) in their analysis that the hypothesis that non-energy commodities, included in the S&P GSCI and in the DJ UBSCI index, are more correlated with oil than the commodity off-index. Their basic assumption is that 'other participants in commodity markets, such as traditional speculators and commercial hedgers, have a limited capacity to absorb trades by index investors. As a result, the growing presence of index investors can affect commodity prices'. Their analysis shows that the average correlation between indexed commodity was indistinguishable from that of the off-index commodity, until the beginning of the new millennium. Although, the lack of a structural analysis does not allow them to identify the main factors of market shock, they likely consider that the expectation of an increase in hedge funds trading contributed to the peak rate of 2007–2008. Then, although hedge funds weren't the causes of price increases, in recent years, they do not exclude their role, while marginal, in relation of these events.

3 Analysis

In addition to the literature analysis—we made a first attempt, albeit perfectible—to identify what statistical relevant quantities can bring out the degree of correlation between spot and future prices, as well as between the different markets. In this regard, 4 markets were considered: US, English, French and Chinese. About the US market, the WEATTKR Index, related to the wheat spot prices at the Kansas City stock exchange, and the GXGRWPSP Index, related to the wheat future prices at the CBOT, are the indices considered. With respect for the English, French and Chinese future markets, they were regarded the LIFFE for the first, the MATIF for the second and the INFAWHC Index for the third.

All prices are for the last ten years, from 2007 to the end of 2016, as the aim of the work is to investigate the fluctuations that took place in the years 2007–2016.

---

[5]The regression model Ordinary Least Squares aims to minimize the sum of residues, i.e. the difference between the observed and the estimated value, to the square.

Furthermore, the market that offers the greatest possibility of analysis is the US, because it is the most liquid given the considerable transaction volume. Conversely, the other markets analyzed are thinner and the data is not always updated. This brief discussion must be concluded by specifying that in the analysis spot prices reported to the Asian market have been overlooked because they showed the same values for an extended period, suggesting the intervention of government authorities in their determination.

At this point, the trend analysis that prices have had over the years and the study of the factors that influenced them in different ways and with different intensity can be started.

From Fig. 2, we note that between 2007 and 2008 there has been a sharp percentage increase in grain prices. These have continued to undergo marked variations even in the following years. There are no high variances between spot and future prices that seem to follow the same trend. Particularly, prices underwent wide and rapid changes between December 2007 and April 2008. In effect, in this period the prices, in addition to undergo lively variations, have reached very high values: the growth began early in the year, with moderate pace and accelerated in September to culminate in the price peaks recorded in the following March, in the spot and future markets. Afterwards, prices started their slight decline, setting on values slightly higher than normal and not undergoing sudden changes until mid-2010. From this point on, they again their rise and variability until mid-2011, when they started to downsize. Furthermore, the previous figure shows that spot prices are much more variable than future prices even if the latter, in the time of high instability in late 2007 to early 2008, showed much more marked values, both increasing and decreasing, going from 192.75 €/T of November 12, to 208.57 €/T of the following week, to go down to the 113.53 €/T of November 26 and then go up again until 251.19 €/T of December 10.

If the horizon is widened and other markets are considered, such as that of the English, French and Chinese derivatives, and are compared to the US market, as in the chart below, we note that the changes are much more pronounced in the latter two countries. Such intense price changes are probably because these are not very thick markets and have fewer transactions than the US market. The greatest changes have occurred in the UK market, while other markets seem to have a very similar trend, although the Asian market is much more volatile than French.

Fig. 2 USA spot and future prices, absolute values. *Source* Bloomberg

Furthermore, American, English and French market prices amounted to similar values and show the same trend, on the other hand, the Asian market takes much higher values, almost twice that of other markets, and even though it follows the trend it accentuates its variability, as can be seen from Fig. 3.

At this point it is appropriate to extend the time horizon to read the current prices in a historical perspective. As shown in Fig. 4, price fluctuations are important compared to the past, but are already known. The increases and fluctuations, in recent years, are not as unusual as would be on a shorter horizon.

The rise of grain prices since 2008 looks amazing. Until the beginning of the new millennium there were no significant increases in prices, while in 2008 and in 2011 they giddily raised. However, from a historical perspective, the price increase and fluctuations in recent years cannot be considered so extraordinary. As can be seen from the previous figure, wheat showed greater fluctuations in the years '70 when the volatility was twice compared to today and changes were not caused by

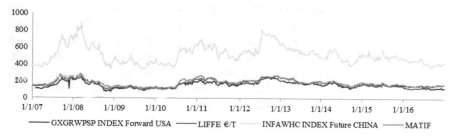

Fig. 3 Wheat prices in USA, UK, France and China, absolute values. *Source* Bloomberg and AHDB

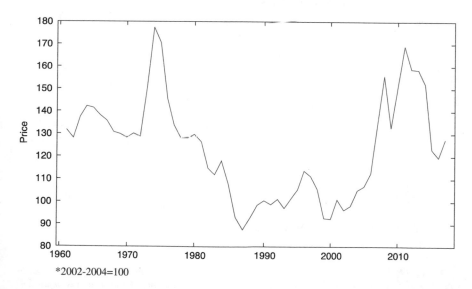

*2002-2004=100

Fig. 4 Food price index. *2002–2004 = 100. *Source* FAOSTAT

deregulated futures markets or by speculative financial transactions, which at that time had certainly not reached their current size.[6]

4 Factors Determining the Trend in Prices

As for all markets, the agricultural sector supply and demand contribute to pricing. Then, the factors that led to the increase in agricultural prices, during the economic crisis, are manifold. Regarding the agricultural products availability product quality and storage are to be considered: indeed, a lower level of supply, a greater quality or a higher inventory level determine price increases, with equal demand. On the supply and demand side, it is the yield that determines the production levels, but unlike sowing, exogenous non-economic factors (Fig. 5).

Over the years, the production level has steadily increased, while the extension of agricultural arable land has been swinging: it underwent a drastic reduction from the years '90 compared to the level reached during the '70s–'80s. The reasons for the decreases in cultivated areas are multiple and attributable to desertification, over-building, industrialization, abandonment of agriculture in order to find more profitable employment in urban areas—to which the problem of agricultural profitability is linked—and, finally, to the mechanization that has led the abandonment of many awkward areas, otherwise arable land even with much more difficulty and physical

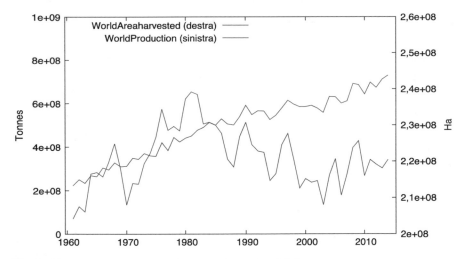

Fig. 5 World production and area harvested. *Source* FAOSTAT

---

[6]Although the origin of derivatives instruments dates back centuries, the birth and the strong developments of financial derivatives can be traced back to the beginning of the seventies of the last century with the introduction of currencies futures which followed, in subsequent decades, the derivatives on interest rates, on stock markets and on credit institutions.

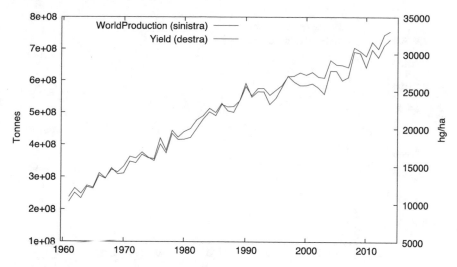

Fig. 6 World production and yield. *Source* FAOSTAT

effort. Accordingly, the highest level of production is attributable to the greater crops yield. In fact, Fig. 6 shows how the yield trend is constantly growing.

Moreover, some academics argue that the rise in prices is driven by a lower land productivity growth than the world's rising population, following the Malthusian[7] postulate of marginal productivity decrease. According to Goodwin et al. (2012), shocks in yields and differences in cultivated areas can contribute to global price instability.

Population growth and well-being and the progress of the middle class have increased food needs and consumption of products with a higher fat content. This means greater fodder production which is in direct competition with food production. This leads to greater pressure from the demand side and, consequently, an increase in prices. Furthermore, the economic development of China and India and the related increase in food consumption impacts heavily on increased demand (Fig. 7).

Another important factor is the climate, given that weather and natural adverse events, such as those that are occurring in recent years—with excessively cold or hot seasons and too much rain and hurricanes—strongly affects the level of production, quality and, consequently, price.

---

[7]In him *An essay on the principle of population and other writings* of 1798, Malthus claimed that mankind had a destiny to hardships and misery if humanity itself does not put a brake on demographic growth. By examining some statistical data, he concluded that population growth was much greater than the ability to produce livelihoods. The fist was increasing, every 25 years, following a geometric progression, while food resources increased more slowly following arithmetic progression. In the face of such imbalance, caused by an excess of population, famines, epidemics and wars were unavoidable events. He called them as a 'positive brakes' because, rising considerably the mortality rate, made food resources again sufficient for the remaining population.

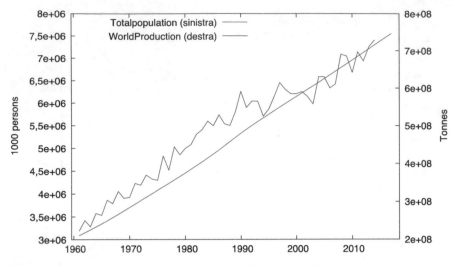

Fig. 7 World total population and world production. *Source* FAOSTAT

The political[8] and macroeconomic environments are equally important because expansionary or restrictive monetary policies may affect storage levels of agricultural commodities, increasing scarcity and prices. Moreover, trade restrictions,[9] determining a degree of internationalization, affect the volatility of a given commodity market. Again, during political upheavals in Crimea,[10] the price of wheat has risen by 27%. And it is well known that the Chinese government, between 2007 and 2012, has used its financial power to secure suppliers of raw materials, oil, offering to Russia, Brazil and other countries million dollars lending.

The oil price is also very important because usually its increase results in a rise in food prices, allowing to highlight a similarity between the curve of crude oil prices and that of other goods, as shown in the following figure. Indeed, putting in the same graph the historical trend of the S&P GSCI Wheat and S&P GSCI Crude Oil, it is clear they have a similar course. At increases of the S&P Crude Oil correspond the increases of the S&P GSCI Wheat; the same dynamic occurs for the decreases. Then, Fig. 8 shows the same relationship, even if the fluctuation of the first are more pronounced than those of the second.

The spread of biofuels, increased in recent times, becomes an alternative to use of agricultural land for food production: hence, recent years have seen a growing increase in cropland to produce biomass for the renewable energy sector such as

---

[8]For example, Argentina's policy to eliminate grain export duties.

[9]'Any trade restriction might harm the match of world supply and world demand and so provoke large price swings. The degree of market internationalization can be measured by the ratio of total world export divided by total world production (in volume). The degree of competition in a given market can also impinge on commodity price movements. Trade restrictions of a given country have a real effect only in case of strong market power' (Geman 2015).

[10]Russia and Ukraine are among the largest producers and exporters of wheat.

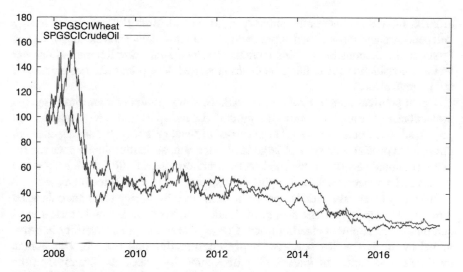

Fig. 8 S&P GSCI wheat and S&P GSCI crude oil. *Source* us.spindices.com

biofuel and biogas, but not to produce food. This happened with the support of public authorities. According to the OECD-FAO (2008), energy and environmental public policies in Europe and the US are the main cause of volatility and price spikes in agricultural commodity markets (Fig. 9).

The graph of biodiesel demand, for Italy and UK, shows very low levels, which tend to be stable and not very significant. Instead, in the US there has been a considerable increase in demand. In particular, except for a significant drop recorded

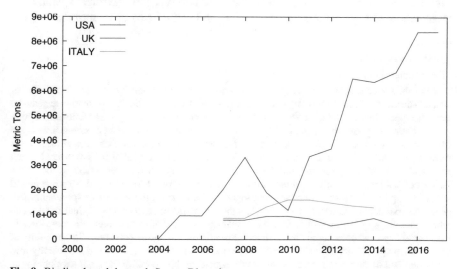

Fig. 9 Biodiesel total demand. *Source* Bloomberg

in 2010, the US shows steady and very rapid growth. Both the policies of energy independence and those aimed at protecting the environment, such as the 2011 Paris Protocol, are responsible for this increase. Exchange rate development is another factor to consider, as the dollar growth has increased flour prices and has curbed the fall in grain price.

About political intervention on the markets, many producer countries triggered protectionist measures to control the production, transport and sale of agricultural products.[11] In Africa, China has been accused of 'land grabbing' for having ensured, through the purchase or lease of large land extension, agricultural resources to protect its national security in the food sector. Furthermore, in 2007, both Russia and Argentina have decreased food exports in response to the local price increase.[12]

Inventory level reduction. Prices can be influenced by storing raw materials when these are low and otherwise increasing trade. The spot market and storage level combine to determine a clearing price. Consequently, the price volatility is determined by the interaction between supply, consumption and storage. The storage level determines most of the volatility of commodity prices, as illustrated in the theory of speculative storage.[13] In fact, low levels of stocks cannot act as a buffer in years of poor harvest. Conversely, high stock help to control supply and demand shock: the warehouse can reduce price fluctuations. Unlike metal and fossil fuel, there are no 'underground reserves' of agricultural products that, every year, are planted, harvested and, in part, wear out. What is not consumed is stored. This relationship between storage and consumption, 'stock-to-use', is very important because the storage absorbs short-term shocks in demand and, especially for agricultural products, protects against negative shocks in supply often caused by adverse weather conditions (Fig. 10).

Arbitrage reflects the influence of agricultural policies; it helps to stabilize price fluctuations and avoids strong peaks. There are two types of arbitrage: spatial arbitrage, where the space factor is used to buy and sell on two different markets; and

---

[11] www.edeport.wur.nl.

[12] www.reuters.com.

[13] In the theory of speculative storage price dynamics is determined by 'optimal' reactions of agents—farmers and those who carry out the storage—who use all available information to generate rational expectations and take excellent positions for purchase and sale of goods. Under assumption of such behavior storage has a stabilizing effect, and the continuing price fluctuations are caused only by repeated random shocks. As Deaton and Laroque (1996) show, in the absence of storage, the prices are a linear fraction of individual shocks: 'The presence of risk-neutral and profit-maximizing stockholders means that expected future prices cannot be greater than current prices by more than the cost of holding inventories into the future, so that, whenever stocks are held from one period to the next, prices in those periods are tied together. As a result, the actions of speculators will generate serial dependence in prices even when there is no dependence in the original shocks, and more generally, speculation will modify the serial dependence that would otherwise come directly from serial dependence in supply and demand. Speculative storage also changes the variability of prices; when speculators hold stocks and their expectation of the future price is sufficiently low, their sales will moderate the effect on price of negative supply (positive demand) shocks.' In the absence of storage, prices would be too low, storage increase them and the sale of stocks, when prices are high, moderates the rise. Then, prices tend to be more stable.

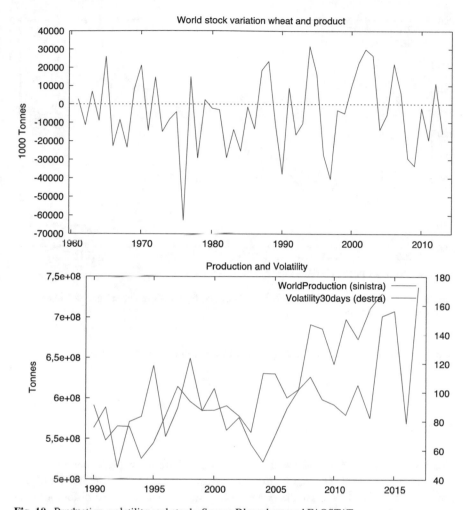

Fig. 10 Production, volatility and stock. *Source* Bloomberg and FAOSTAT

temporal arbitrage, in which one acts on the time factor with the sale after purchase. In the case of wheat, temporal arbitrage is negatively correlated with the final stocks (Prehen et al. 2014). Conversely, regarding space arbitrage there is a correlation between volatility and exports: an increase of the second is an increase of the first, but this relationship has not yet been completely clarified.

Buyuksahin and Robe (2014) show that the correlation between commodity yields and stock market indices grows if there is a greater participation of speculators and hedge funds, especially those who have open positions in both the equity and commodity futures markets. However, as shown in the following figure, after ten years when commodity prices moved in unison with the equity market, recently the

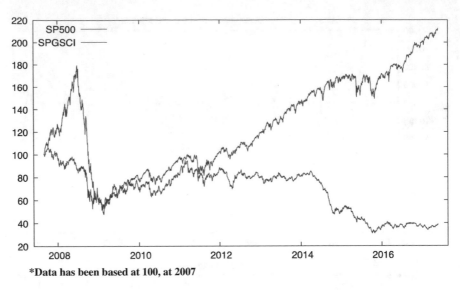

*Data has been based at 100, at 2007

Fig. 11 Correlation between S&P 500 and GSCI commodity. *Data has been based at 100, at 2007. *Source* us.spindices.com

two markets have taken different paths: the correlation from 2016 is at the lowest (Fig. 11).

5 Conclusion

Grain prices from 1960 show a stable growth trend with a few peaks between years '70–'80 and in 2008. Despite a strong decrease, prices are still higher than pre-financial crisis levels and are characterized by high volatility. Several factors help explain market instability. Food safety and the sustainability of food resources are closely linked to political stability.[14] Moreover, agriculture is at the heart of the problems of global population growth, of soil erosion, of shortage of arable land, of trade financialization, of competition for land with mining and urban development, as well as of competition for water, an increasingly important and scarce resource.[15] Then, the search for sustainability indicates that economic growth is possible only within social development, which, in its turn, must be based on respect for and enhancement of the environment. In the achievement of sustainability, the role played

---

[14]States that cannot safe and economic access to food for their populations are much more likely to face protests and suffer political instability. For some, the Arab Spring, which began in 2011, is an excellent example of all this. For the Arab word, in fact, the food price was an important reason for the 2011 protests.

[15]For example, the international tensions concerning water between India and Pakistan.

by financial and real markets and by their operators, identified in financial, credit or commercial intermediaries, is equally important.

Numerous studies that have put speculation at the base of high food prices have been denied by many academic studies. Climate and country risks are the two main factors behind high prices.

It is not possible to clearly highlight a direct link between speculation, price growth and market volatility: this is also because the presence of professionals favors a better functioning of the market—incensing liquidity, to the point that the lack of speculation would put the very existence of the market at risk.

Futures markets, often blamed for increased volatility, can be an essential element of risk management in physical markets, because they must be able to provide an infrastructure that always guarantees a market price. Jacks (2007) investigates the relationship between futures markets, speculation and commodity prices volatility and shows that futures markets are systematically associated with lower levels of price volatility.

Despite speculation, commodity trading has been regularly denoted as responsible for high prices and high volatility, especially by governments facing a difficult situation in their country. It is important to keep in mind that in August 2013 the total value of 'open interest' on agricultural futures was 9.6% of the world production of one year and 3% of the individual annual transactions of all physical markets: therefore, this is a negligible percentage. If we compare the Gxgrwpsp Index with the 'speculative variable' contained in the CFTC reports such as Open Interest that shows the total of all contracts entered into and not yet offset at the date of reference; Commercial and Non Commercial long positions—we have the first ones if the trader use futures contracts in that particular commodity for hedging otherwise we have the latter; CIT—Commodity Index Trader—long positions, category entered in 2006 by CBOT to indicate speculative indices on commodities, and the total Reportable and Non Reportable positions,[16] we note that future prices at the beginning of an investigation period—early 2007—had a reverse trend compared to other 'speculative variables'. After this period, when we hear speculation and prices begin to increase, the two lines follow the same trend accentuating or dampening the fluctuations. However, in 2016, when prices mitigated their volatility they are divided: prices decrease, and positions increase. Only exception is for the CIT that— although in early 2007 —show an opposite trend compared to prices. In 2012 they begin to follow the trend accentuating the fluctuations, and continue in the following period until the end of 2016, although dampening the fluctuation, thus diverging from the performance of the other 'speculative' variables. The cointegration analysis it is necessary to show that there is no evidence that financial derivative instruments determine the fluctuations of wheat futures price.

Will et al. (2012) analyzing the influential of financial speculation on agricultural commodities, conclude that 'according to the current state of research, there is little supporting evidence that the recent increase in financial speculation has caused either (a) the price level or (b) the price volatility in agricultural markets to rise'.

---

[16]*Source* CFTC.

References

Adammer, P., Bohl, M. T., & Gross, C. (2016). Price discovery in thinly traded futures markets: How thin is too thin? *Journal of Futures Markets, 36*(9), 851–869.

Baldi, L., Peri, M., & Vandone, D. (2013). Price discovery in commodity markets. *Applied Economics Letters, 20*(4), 397–403.

Bauer, P., & Minsch, R. (2013). Commercio di materie prime agricole: croce o delizia? *Dossierpolitica Economiesuisse, 4.*

Brunetti, C., Buyuksahin, B., & Harris, J. H. (2015). *Speculators, prices, and market volatility.* Working paper. Bank of Canada.

Buyuksahin, B., & Robe, M. A. (2014). Speculators, commodities and cross-market linkages. *Journal of International Money and Finance, 42,* 38–70.

Chevallier, J., & Ielpo, F. (2013). Cross-market linkages between commodities, stocks and bonds. *Applied Economics Letters, 20*(10), 1008–1018.

Deaton, A., & Laroque, G. (1996). Competitive storage and commodity price dynamics. *Journal of Political Economy, 104*(5), 896–923.

Geman, H. (2015). *Agricultural finance. From crops to land, water and infrastructure.* UK: Wiley Finance Series.

Gilbert, C. L. (1995). Modelling market fundamentals: a model of aluminum market. *Journal of Applied Econometrics, 10*(4), 385–410.

Gilbert, C. L. (2010). How to understand high food prices. *Journal of Agricultural Economies, 61*(2), 398–425.

Goodwin, B. K., Marra, M., Piggot, N., & Mueller, S. (2012). *Is yield endogenous to price? An empirical evaluation of inter and intra-seasonal corn yield response.* North Carolina State University.

Hamilton, J. D., & Wu, J. C. (2015). Effects of index-fund investing on commodity futures prices. *International Economic Review, 56*(1), 187–205.

Irwin, S. H., & Sanders, D. R. (2010). The impact of index and swap funds on commodity futures markets: Preliminary results. In *OECD food, agriculture and fisheries.* Working papers no. 27.

Ivanov, I. (2011). Cross-sectional analysis of index and commodity markets price discovery. *Global Business and Finance Review, 16*(2), 1–16.

Jacks, D. S. (2007). Populist versus theorist: Futures markets and the volatility of prices. *Exploration in Economic History, 44*(2), 342–362.

Kim, A. (2015). Does futures speculation destabilize commodity markets? *Journal of Futures Market, 35*(8), 696–714.

Master, M. W. (2009). *Testimony before the committee on homeland security and government affairs.* US Senate.

Master, M. W., White, A. K. (2008). *The accidental hunt brothers: How institutional investor are driving up food and energy prices.* Special report 2008.

OECD-FAO. (2008). *OECD–FAO agricultural outlook 2008–2017.*

Ott, H. (2014). Extent and possible causes of intrayear agricultural commodity price volatility. *Agricultural Economics, 45*(2), 225–252.

Prehen, S., Glauben, T., Loy, J. P., Pie, I., & Will, M. G. (2014). *The impact of long-only index funds on price discovery and market performance in agricultural futures market.* Discussion paper no. 146, Leibniz Institute of Agricultural Development in Transition Economies.

Singleton, K. J. (2014). Investor flows and the 2008 boom/bust in oil prices. *Management Science, 60*(2), 300–318.

Tang, K., & Xiong, W. (2012). Index investment and the finanzialization of commodities. *Financial Analyst Journal, 68*(6), 54–74.

Will, M. G., Prehn, S., Pies, I., & Glauben, T. (2012) *Is financial speculation with agricultural commodities harmful of helpful?—A literature review of current empirical research.* Discussion paper no. 2012–27. Martin-Luther-University Halle-Wittenberg.

Financial and Non-financial Reporting: Examining the Past to Understand the Future

Maria Serena Chiucchi, Marco Giuliani, Simone Poli, Marco Gatti, Marco Montemari and Luca Del Bene

Abstract Nowadays, traditional sources of wealth are becoming more and more intangible (i.e., knowledge, customer relationships, etc.). The traditional accounting systems are no longer considered able to adequately support the management (see strategic or managerial control/reporting systems) and the disclosure (see financial statements) of intangibles and, indeed, of the whole organisation. The studies developed by the Department of Management concern both financial and non-financial disclosure. Regarding the former, research is focused on whether and how it would be possible to improve the quality of the disclosure of intangibles, leveraging on the degree of freedom provided by the law and national and international accounting standards. As regards the latter, research is focused both on the reporting of intangibles/intellectual capital and, more recently, on integrated reporting frameworks and on the assurance of non-financial information.

M. S. Chiucchi (✉) · M. Giuliani · S. Poli · M. Gatti · M. Montemari · L. Del Bene
Department of Management, Università Politecnica delle Marche, Ancona, Italy
e-mail: m.s.chiucchi@univpm.it

M. Giuliani
e-mail: m.giuliani@univpm.it

S. Poli
e-mail: s.poli@univpm.it

M. Gatti
e-mail: m.gatti@univpm.it

M. Montemari
e-mail: m.montemari@univpm.it

L. Del Bene
e-mail: l.delbene@univpm.it

© Springer Nature Switzerland AG 2019
S. Longhi et al. (eds.), *The First Outstanding 50 Years
of "Università Politecnica delle Marche"*,
https://doi.org/10.1007/978-3-030-33879-4_12

1   Introduction

Since the 1980s companies have been called to face a radical change in the financial and commodity markets and their competitive dynamics. In particular, the increasing importance of key success factors like innovation, customer satisfaction, personnel competencies, product and process quality, timeliness, information management, etc., has fostered the birth of ideas and concepts such as *knowledge economy, information economy* and *network economy* where knowledge, information and relationships are considered necessary elements for survival. Accordingly, companies have changed their business models, their strategic processes and their way of interacting with their stakeholders. In sum, over the last 20 years, both the markets and companies had to undergo a *revolution*, i.e. a shift from an economy based on tangible items to a one focused on intangibles.

This revolution also impacted accounting studies. In fact, one of the most profound effects was that the most widespread tools used by companies for internal (managerial) or external (disclosure) purposes, ones predominantly focused on financial information, were unable to fully capture the value of the abovementioned *new* sources of value creation. More in-depth, financial statements were being criticized as they were not able to fully account for a large part of a firm's intangibles or for their performance; this has led (and it is still leading) to a significant gap between market values and book values and, consequently, to a relevant loss of financial accounts. Similarly, measurement of organisational performance for managerial purposes through financial indicators alone was considered inadequate support for managers making value creation-oriented decisions, because numerous strategic resources and activities cannot be represented to their full extent solely in monetary terms. In short, the need to *make the invisible visible* has arisen and, consequently, new accounting concepts, methods, and tools have been proposed.

Moving on from these considerations, we look at how scholars, practitioners, and policymakers have developed their activities along two main pathways. The first adopts a value realisation perspective, and it is related to the widening of the boundaries of financial statements and to how it is possible to report the financial value of a firm's reliably and to provide investors with more useful information. The second, instead, is focused on the idea of value creation and on the design of reporting methods and tools useful for measuring and disclosing the stocks and flows related to the organization's non-financial capital, such as its human, relational, organisational, natural, or social capital, that cannot be represented adequately in financial statements. As evidenced in extant studies, these two research streams have been developed considering business reporting both from a production perspective and from a consumption one, according to the phases of the development process of a measurement system (Bourne et al. 2000; Ferreira and Otley 2009; Neely and Bourne 2000; Roslender and Fincham 2004). The production perspective regards the design and implementation of the reporting model, while the consumption perspective concerns the actual application and use of the designed and implemented system by company actors to obtain the expected outcomes.

This chapter aims to present the latest scientific contributions offered by the research conducted by the Department of Management scholars in the fields of financial and non-financial reporting; in addition, future research avenues will be proposed.

The structure of this chapter is the following. The next section offers an overview of past, current, and future research in the field of financial reporting. Section 3 presents the state of the art and the work in progress in the field of non-financial reporting. The last section summarises the main insights gleaned and draws some conclusions.

2 Financial Reporting: Past, Present, and Future

Financial reporting is probably the most traditional and relevant area of research in the field of accounting; consequently, financial statements and interim reports have been the object of a plethora of national and international publications.

According to the International Accounting Standards Board (IASB) framework and the Italian GAAP, the objective of financial statements is to provide information about the financial position, performance, and changes in the financial position of an enterprise that is useful to a wide range of users that need to make economic and/or financial decisions. In other words, the idea underlying the practice of financial disclosure is to provide useful information to the stakeholders of a firm (e.g. investors, partners, suppliers, customers) about the financial performance and the value of the firm itself.

Due to the evolution of the competitive arena and to the internationalisation and globalisation processes, the standard setters have defined and implemented a rich agenda in the aim of defining accounting standards able to improve the usefulness of financial statements.

Delving into the changes brought to accounting standards can be approached from different research perspectives, such as analyzing how the accounting standards have been applied in practice, examining the convergence process between the Italian Generally Accepted Accounting Principles (GAAP) and International Financial Reporting Standards (IFRS), investigating the value relevance of financial information, examining the practice of earnings management, and analyzing the use of financial information.

The contributions in the field of financial reporting, made in recent years by scholars of the Department of Management, will be discussed in the following sub-sections.

2.1   Financial Reporting: The State of the Art

The research developed up to now in the field of financial accounting mainly regards the production of financial information from a practice-based perspective.

The first stream of research developed by the scholars of the Department of Management regards the disclosure and valuation of intangibles. Intangibles represent one of the Department's main areas of research and they are analysed in terms of visualisation, measurement and disclosure. More specifically, concerning the valuation of intangibles, the research shows that the value of intangibles tends to present a limited level of objectivity, consistency, comparability and understandability. Nevertheless, the valuation of intangibles is essential, if one is to make the invisible visible; in other words, it can be considered an opportunity to visualise and understand the influence of intangibles on financial performance (Giuliani 2014, 2016a; Giuliani and Marasca 2011; Marasca and Giuliani 2009). The disclosure and use of the information related to intangibles have been examined from several aspects. Research has been focused on how and to what extent intangibles can be disclosed in financial statements (e.g. in the management commentary, in the notes in the balance sheet or the income statement); it has also delved into the influence of culture and national accounting practices on the disclosure and the effects generated by the disclosures (Brännström and Giuliani 2009a, b; Brännström et al. 2009; Giuliani 2003, 2006, 2016a; Giuliani and Marasca 2017, 2018a, b).

The second stream of research concerns the issues of accounting regulation and the gap between *de iure* and *de facto* accounting, i.e. the difference between what companies are expected to do and what they actually do. Due to the Department's cultural roots and the widespread strong belief that scientific research should also have an impact on society, scholars have analysed the practice of accounting to understand what firms do and suggest what they should do to improve the usefulness of financial disclosure. Moreover, it is worth mentioning several studies regarding the application of the Italian GAAP both by industrial, trade, and service companies and by banks and financial companies. These studies have shown that firms are generally compliant with the accounting standards, but they tend to provide as little information as possible and adopt the easiest solution offered by the standards rather than the best one. This implies that the information asymmetry that accounting standards aim to reduce tends, instead, to persist in practice. Furthermore, it emerges that even in case of a change in the accounting standards companies tend to maintain as much as possible the previous approach in order not to change the accounting procedures. In other words, the phenomenon of accounting inertia tends to dominate accounting practice and to impede companies from taking advantage of the disclosure opportunities offered by the new accounting standards (Brännström and Giuliani 2009a; Giuliani and Brännström 2011). In this context, the studies developed with reference to the concept of materiality and how it is applied in practice (Branciari and Poli 2009; Branciari et al. 2007; Poli 2013a) as well as the research on the impairment of tangible assets (Poli 2008, 2012a) and on the impact of taxation rules on financial reporting/accounting choices (Poli 2015a) also bear noting.

Another stream of research regards the practice of earnings management, i.e., a strategy used by the management of a company to deliberately manipulate the company's earnings so that the figures match a pre-determined target. These studies regard, in particular, the so-called *earnings minimization practices* and *earnings change minimization practices* in Italian SMEs. The aim is to understand how widespread these practices are, what their determinants are (e.g. variables related to ownership structure, corporate governance, gender, etc.), and whether they present some specificities in comparison to the ones examined in other studies (Branciari and Poli 2017; Poli 2013b, c, 2015b, 2017a, b).

Auditing has also been the object of attention. Research in this area has regarded the practice of financial auditing, i.e., how auditors perform their activity (what and how they audit) in their everyday life, how policymakers and clients perceive the role of auditors, and how the auditors themselves interpret it. It also investigates the idea of *reciprocal affinity* between clients and auditors (Branciari and Giuliani 2005; De Santis and Giuliani 2014, 2015; Giuliani 2010).

Accounting history has also been investigated in the last decade. In this area, the focus has been on how accounting technologies have been applied in the past to induce specific behaviours or support specific managerial or governmental practices (Gatti and Poli 2011, 2014, 2018; Poli 2011, 2012b).

In summary, the studies presented in recent years have undertaken to analyse the different significant dimensions of financial reporting using both quantitative and qualitative research methods, in the aim of taking an active role in the international and national scientific debate.

2.2 Financial Reporting: Future Research Avenues

Recently, several national and international accounting laws have been modified to improve the quality of financial statements and carry out the accounting harmonisation process, i.e., a convergence process among national and international accounting settings. Consequently, accounting standards have been deeply revised.

In Italy, Decree 139/2015 eliminated many of the previously existing differences with IAS/IFRS. For example, the notion of fair value was introduced as a method for recognising and measuring derivative financial instruments and that of amortised cost as a criterion for measuring receivables, payables, and securities. In addition, financial statements are required to provide greater emphasis to related parties' transactions, to cash flow statements, to the substance over form principle, and to a consistent presentation of hedging transactions. The reason for all these innovations is to allow transparent, qualified, and up-to-date information to be provided, information that is suitable for expressing the status of investments and sources of funds and the company's ability to raise and absorb financial resources. In sum, this will improve the transparency of financial information and facilitate access to financial markets for Small and Medium-sized Enterprises (SMEs). In addition, the IASB has defined a rich working agenda, the purpose being to update the extant accounting standards IAS/IFRS and regulate the accounting of new business phenomena. Also in this case, the underlying goal is to improve the quality of financial reporting and, consequently, its usefulness for investors, in the hopes of, ultimately, increasing the effectiveness and efficacy of financial markets.

Moving forward, the Department's research will be focused on understanding the impact of the new accounting standards on organisational processes and the dynamics of financial markets. In addition, as the option of adopting international accounting standards instead of the national ones was introduced ten years ago, it is now possible to develop a diachronic analysis in order to understand the concrete internal (organizational) and external (market related) impacts of the adoption of international accounting standards in Italy on a mandatory or voluntary basis. In developing these studies, particular attention will be devoted to the emerging accounting and valuation practices related to intangibles and organizational non-financial capital.

Another recent reform concerns bankruptcy and insolvency procedures in Italy (Law 155/2017). Seeking to ensure the best interest and satisfaction of creditors, the New Code prioritises procedures aimed at overcoming the crisis by keeping the business as a going concern (even if under new ownership). The new measures introduced by Law 155/2017 include: (a) an alert system (*procedura di allerta*) based on an early warning procedure to be exercised on a confidential basis and completed within reasonably short timeframes, to improve the efficiency and speed of reaction to a financial crisis; (b) out-of-court crisis settlement procedures to ensure early disclosure of economic distress and facilitate negotiations with the distressed company's creditors; (c) incentives for business continuity; (d) provisions aimed at coordinating the management of a crisis affecting a group of companies; (e) incentives for the early disclosure of facts or events that are reasonable indicia of distress or insolvency; (f) an increase in the duties and responsibility of internal and external auditors with reference to situations of potential or actual financial distress.

Several investigations have been conducted to develop bankruptcy prediction models (Altman and Hotchkiss 2010; Bellovary et al. 2007; Fejér-Király 2015). Although they are generally considered to be efficient and effective, the current prediction models present some criticalities such as: (a) the excessive focus on accounting data and the lack of forward-looking information; (b) the focus on American large (listed) firms; (c) the lack of industry-specific or country-specific dimensions; etc. Several scholars and practitioners have performed studies referred to the Italian context, but they present and recognize some relevant scientific and practical limitations such as the identification of a very large area of uncertainty and the lack of adequate background or validation procedures. In summary, there is the need to develop a large-scale study to reconsider the existing prediction models in order to adapt them to the Italian context and to make them compliant with the requisites indicated by the law.

The scholars in the Department of Management will carry out research that aims to analyse the accuracy and effectiveness of bankruptcy prediction models concerning Italian SMEs, given that the extant studies are mainly referred to large American firms. Moreover, studies regarding the quality and effectiveness of restructuring business plans will be developed, considering both the perspective of the preparers and of the users. Finally, investigations regarding the earnings management practices of financially distressed companies will also be carried out.

Another important accounting reform is the one that has changed the auditing regulation. In particular, the reform (Decree 39/2010) made the adoption of the International Standards of Auditing obligatory, and it made deep changes to the profession of auditor in order to bring Italian auditing activity into alignment with international auditing and increase the quality of auditing procedures. Decree 135/2016 modified the rules introduced by Law 39/2010 in order to make the Italian rules compliant with the European regulations and the new auditing standards. This reform offers the opportunity to understand to what extent the auditing profession has changed in practice and the effects of the auditing reform on the dynamics of financial markets.

The studies on accounting history will continue following the research lines that have been at the core of research conducted by the Department in recent years.

3 Non-financial Reporting: Past, Present, and Future

3.1 Non-financial Reporting: The State of the Art

Similarly to the financial reporting discourse, the non-financial one has also been influenced by the change of context outlined above. In fact, over the last 20 years, several new concepts, methods, and tools have been proposed in order to improve non-financial reporting for internal or external purposes, in the aim of better supporting the managerial decision process and of offering the stakeholders a complete representation of the organisational value creation process.

Within the field of non-financial reporting, the research developed to date by Department of Management scholars regards both the production and consumption of reports for internal and external aims. This research has mainly adopted a practice-based perspective, as done for the financial reporting field.

Regarding non-financial reporting for managerial purposes, a first stream of research was devoted to the analysis of the design, implementation, and use of management accounting tools able to ensure a broader view of company performance. Particular attention was devoted to strategic performance measurement tools, like the Balanced Scorecard, showing how they can ensure a multidimensional measurement of a company's performance and highlighting the costs and benefits which may arise when they are implemented in different contexts, for-profit and also non-profit (Gatti 2011a, 2015; Gatti and Chiucchi 2017; Marasca and Chiucchi 2008; Marasca and Del Bene 2009; Montemari 2018a). Moreover, a large part of the research was focused on the way financial and non-financial information could and should be combined in order to support managers' decision-making processes (Gatti 2011b, 2013) as well as on analyzing how information technology tools and management accounting systems interact and influence each other, thereby creating synergies which favour the production and consumption of non-financial and multidimensional information (Chiucchi et al. 2012; Del Bene and Ceccarelli 2016; Nespeca and Chiucchi 2018).

Parallel to the studies about multidimensional measurement systems, other studies have focused on a specific dimension of organisational performance, i.e., intangibles. Indeed, this area can be considered the main research stream pursued by the Department of Management. Under the term intangibles, research paths also focus on intellectual capital, intended as the whole system of a company's intangible resources.

Accounting studies about intellectual capital have developed over time, following an ideal path made up of three stages: (1) raising awareness on intellectual capital relevance and creating frameworks to measure it; (2) gathering empirical evidence and building intellectual capital theories; (3) developing a critical and performative analysis of intellectual capital in action. The Department of Management has followed this path and developed research that contributed to each of the stages. We now take a more in-depth looks at the progression of Department research.

At first, attention was focused on the design of measurement tools that could deliver external and internal reports, i.e., intellectual capital reports, able to supplement information included in the financial statement and provided by traditional (financial) performance measurement systems (Chiucchi 2004, 2005, 2008; Chiucchi and Marasca 2004). Responding also to the call for research approaches which could favour the osmosis between theory and practice and fill the gap between these two realms, an interventionist research approach and, more specifically, a constructive research approach (Chiucchi 2012; Jönsson and Lukka 2007) was adopted. This choice was made as it allows for the development of *innovative constructions* (e.g. intellectual capital measurement systems, intellectual capital reports) geared to solving real-world problems (*practical relevance*) to which no theoretical solution had previously been found, and able to also contribute to theory advancements (*theoretical relevance*). Using this approach also requires the implementation of the proposed solution to test its applicability (*practical functioning*). By focusing on the production of intellectual capital reports, the Department of Management scholars contributed to the first two stages of the intellectual capital discourse.

Subsequently, attention turned to the actual use of the designed and implemented tools in order to explore, from a critical and performative perspective, what happens in companies when these tools are designed and implemented; the studies focused on the organisational and behavioural dynamics triggered by the design and use of these reporting tools (Chiucchi 2013a; Chiucchi and Dumay 2015). The research that followed aimed to provide insights on what works and does not work when it comes to actually using these tools (Chiucchi et al. 2014, 2016), on how companies make sense of and give sense to intellectual capital indicators (Giuliani 2016b; Giuliani et al. 2016), on the levers and the barriers that can enable or hinder the use of these tools (Chiucchi 2013b; Chiucchi and Montemari 2016; Chiucchi et al. 2018a; Montemari and Chiucchi 2013), on the role played by the different actors involved in the reporting process (Chiucchi and Giuliani 2017; Chiucchi et al. 2018b; Giuliani and Chiucchi, in press), and on how barriers can be overcome (Montemari and Chiucchi 2017). Moreover, it is worthy of mention that much of the performative research was focused on the dynamic aspects of intellectual capital in order to explore how intellectual capital elements combine with each other for the sake of value creation (Montemari and Nielsen 2013, 2014) and how these elements change over time (Giuliani 2015; Giuliani and Skoog 2017). Through these studies, several contributions based on the empirics of practice have been presented to foster a deeper understanding of what intellectual capital reporting does in organisations and markets.

With reference to non-financial reporting for external purposes, the Department of Management scholars have developed research devoted to the analysis of new ways of disclosing corporate performance through voluntary reports, such as social reports, intellectual capital reports and, more recently, integrated reports. In this field, the research has been focused on how organizations have applied the most widespread social accountability standards (e.g. AccountAbility1000, AA1000SES, Italian standard for social accountability in healthcare organisations, and Global Reporting Initiative Framework) (D'Andrea 2012, 2017; Marasca et al. 2018; Montanini and D'Andrea 2012) and followed the most well-known intellectual capital reporting guidelines and models (e.g. DATI, Meritum, etc.) (Chiucchi 2008, 2013b; Chiucchi et al. 2016). In this research arena, too, the research has followed the process of the design and implementation of social reports. The findings have mainly underlined the adjustments needed to apply the different frameworks and to produce non financial reports as external communication tools, the barriers, and the challenges to its practical implementation in specific contexts.

More recently, attention has been centred on integrated reports (especially in the version proposed by the International Integrated Reporting Council—IIRC) which can be understood as the natural evolution of management accounting tools and research in the field of intellectual capital, social reports, and business models. In fact, integrated reporting is aimed at integrating financial and non-financial information in an attempt to communicate not only the value created by a company but the way it is created, in particular. From this standpoint, integrated reporting research can be considered a synthesis of previous research; at the same time, it has opened another very prolific avenue of research. Studies have been primarily oriented towards the process of design and implementation, highlighting the adjustments needed to apply the IIRC frameworks, the barriers, and the challenges to its practical implementation (Marasca et al. 2017). Subsequently, attention was devoted to the way existing management accounting systems can support the adoption of these new reporting tools and, in a different perspective, to how their adoption can affect management accounting systems and lead to a change of the nature of the information provided or of the role played by the management accountant (Chiucchi et al. 2018a, b; Gatti et al. 2018). The research was conducted in different contexts, to provide a broad overview of the different ways in which new reporting tools can influence, or can be influenced by, management accounting systems.

As an ideal bridge between internal and external reporting, studies about business models have also grown in recent years. The research on business models, i.e., the platform through which companies create, deliver, and capture value, has gained relevance through a combination of conceptual and empirical studies focused on the contribution of business models to performance measurement (Nielsen and

Montemari 2012; Montemari and Chiucchi 2017) and on business model innovation (Lüttgens and Montemari 2016; Taran et al. 2016). Moreover, attention has also been devoted to the analysis of how the business model research area has developed over time in order to provide a guide for future research and theorisation (Lambert and Montemari 2017; Montemari 2018b; Nielsen et al. 2018a, b).

It must be noted that the rise of new tools to measure company performance has also stimulated a debate about the methodological approaches through which their adoption and use should be analysed and studied. In this context, the research activity was focused on the contribution that the qualitative approach and the case study method, in particular, can offer to the exploration of these new measurement tools and of the way they work in practice (Chiucchi 2012, 2014).

3.2 Non-financial Reporting: Future Research Avenues

Future research in the field of non-financial reporting will proceed along the path drawn in the past, i.e., reporting for managerial and disclosure purposes.

As regards non-financial reporting for managerial purposes, given the huge quantity of research carried out in recent years to explore the full potential of multidimensional measurement systems, it could be useful to deepen the analysis of factors that can make their implementation complex or that, in many cases, can hinder their use (consumption) within companies. This could be done by resorting to interventionist research to fill the gap between theory and practice and favour the osmosis between them and by using an empirical analysis in an attempt to provide useful suggestions to praxis.

Another relevant avenue of research on managerial reporting concerns the contribution that big data and artificial intelligence can give to the functioning of performance measurement systems and the impact they can have on the provision as well as on the use of strategic and non-financial information. In line with this, it would be fruitful to explore how business intelligence systems can promote the adoption and use of performance measurement systems, shedding light on the effects that can arise, in terms of measurement, when these systems are implemented and, at the same time, on the barriers that can limit their actual adoption within organizations. In short, the overall impacts of leading-edge technologies on managerial reporting will be investigated.

As concerns non-financial disclosure, the focus will be on two main types of report: integrated reports and non-financial statements. In fact, the recent EU directive 2014/95/UE as well as Decree 254/2016 have deeply influenced the current reporting practices. Non-financial disclosure is mandatory for some companies, and these regulations implicitly promote the adoption of the reporting standards issued by the Global Reporting Initiative or by the IIRC.

In this context, investigations will be centred on the design and implementation of these reports or statements, with a particular focus on the conditions that could ensure comparability of different reports in space and time and that could enhance the adoption of such reports in peculiar contexts, e.g. SMEs. Furthermore, the use of these kinds of reports will be studied in order to understand the sense making processes they lead to and consequently, how the main organisational stakeholders consume them. Finally, the practice of non-financial assurance will be examined, i.e., how auditors act in order to verify the statements made and the role they play in the non-financial disclosure process.

As mentioned above, research fields that can be considered bridges between internal and external reporting - for instance, business models - will be explored. Considering that business models represent a mature research area, future research will be directed toward providing performative contributions about what happens in companies when business models are designed, implemented, and used. This could be useful to show what works and what does not work, the levers and barriers that can enable or hinder the design of a business model, as well the reasons underlying negative or positive experiences with their use. Additionally, integrated thinking and integrated governance, i.e., the use of integrated reports to support managerial decision-making processes and to promote a holistic managerial approach, also represent bridge areas of research. Future investigations can regard the levers and barriers related to the use of integrated reports for internal purposes as well as the connections between reporting and management.

4 Final Remarks

This chapter aimed to present the most up-to-date scientific contributions offered by the research undertaken by the Department of Management scholars in the fields of financial and non-financial reporting; part of this objective was also to ponder and propose future research avenues.

In the field of financial reporting, the studies have been mainly centred on the practice of accounting, i.e., on how accounting and auditing standards are applied, on earnings management practices, and on how organisations are evaluated. With regard to non-financial reporting, the research has primarily focused on the production and consumption of reports for managerial (such as multi-dimensional performance measurement systems, intellectual capital measurements) or disclosure purposes (such as corporate social responsibility reports, intellectual capital reports, integrated reports, non-financial statements). These studies have mainly adopted an interventionist research approach, and they have involved private and public organisations, especially the ones that can be classified as SMEs as they tend to be overlooked by mainstream research even if they are highly relevant in the Italian and European contexts.

Regarding future research activities, the Department scholars will continue to conduct studies in the fields of both financial and non-financial reporting. More specifically, the goal is to analyse the development of accounting practices as new business phenomena arise and as management and stakeholders' information needs evolve.

In summary, the fil rouge that connects the different studies produced to date and the ones that will be in the future is the vision of carrying out research that is scientifically relevant and able to improve and enhance practice and of bridging the theory-practice gap. In this light, the Department of Management's accounting scholars will continue developing research studies based on the empirics of practice collected in private and public organisations, with a particular focus on SMEs.

The validity of this approach is upheld by the quantity and quality of scholarly publications emanating from the Department of Management, as well as by the quantity and value of the applied research performed to date and, by reasonable assumption, in the near future.

References

Altman, E. I., & Hotchkiss, E. (2010). *Corporate financial distress and bankruptcy: Predict and avoid bankruptcy, analyze and invest in distressed debt*. Hoboken: Wiley.

Bellovary, J. L., Giacomino, D. E., & Akers, M. D. (2007). A review of bankruptcy prediction studies: 1930 to present. *Journal of Financial Education, 33*, 1–42.

Bourne, M., Mills, J., Wilcox, M., et al. (2000). Designing, implementing and updating performance measurement systems. *International Journal of Operations & Production Management, 20*(7), 754–771.

Branciari, S., & Giuliani, M. (2005). Le procedure di controllo del bilancio consolidato: Un'indagine empirica. *Revisione Contabile, 64*, 29–39.

Branciari, S., Giuliani, M., & Poli, S. (2007). Il principio di significatività e rilevanza nella prassi italiana: Alla ricerca di un comportamento condiviso. *Rivista dei Dottori Commercialisti, 58*(6), 1051–1078.

Branciari, S., & Poli, S. (Eds.). (2009). *Il principio di rilevanza nella prassi dei bilanci italiani*. Torino: Giappichelli.

Branciari, S., & Poli, S. (2017). The impact of gender diversity in boards of directors on "earnings minimization" in Italian private companies. *International Journal of Business and Social Science, 8*(10), 130–138.

Brännström, D., Catasús, B., Gröjer, J. E., & Giuliani, M. (2009). Construction of intellectual capital. The case of purchase analysis. *Journal of Human Resource Costing & Accounting, 13*(1), 61–76.

Brännström, D., & Giuliani, M. (2009a). Accounting for intellectual capital: A comparative analysis. *VINE Journal of Information & Knowledge Management, 39*(1), 68–79.

Brännström, D., & Giuliani, M. (2009b). Intellectual capital and IFRS3: A new disclosure opportunity. *Electronic Journal of Knowledge Management, 7*(1), 21–23.

Chiucchi, M. S. (2004). *Sistemi di misurazione e di reporting del capitale intellettuale: Criticità e prospettive*. Torino: Giappichelli.

Chiucchi, M. S. (2005). Measuring intellectual capital in small and medium enterprise: The Iguzzini illuminazione case study. In *Proceedings of the 1st EIASM Workshop on Visualising, Measuring and Managing Intangibles and Intellectual Capital*. Ferrara, October 18–20, 2005.

Chiucchi, M. S. (2008). Exploring the benefits of measuring intellectual capital. The Aimag case study. *Human Systems Management, 27*(3), 217–230.

Chiucchi, M. S. (2012). *Il metodo dello studio di caso nel management accounting*. Torino: Giappichelli.

Chiucchi, M. S. (2013a). Intellectual capital accounting in action: Enhancing learning through interventionist research. *Journal of Intellectual Capital, 14*(1), 48–68.

Chiucchi, M. S. (2013b). Measuring and reporting intellectual capital: Lessons learnt from some interventionist research projects. *Journal of Intellectual Capital, 14*(3), 395–413.

Chiucchi, M. S. (2014). Il gap tra teoria e prassi nel Management Accounting: Il contributo della field-based research. *Management Control, 3,* 5–9.

Chiucchi, M. S., & Dumay, J. C. (2015). Unlocking intellectual capital. *Journal of Intellectual Capital, 16*(2), 305–330.

Chiucchi, M. S., Gatti, M., & Marasca, S. (2012). The relationship between management accounting systems and ERP systems in a medium-sized firm: A bidirectional perspective. *Management Control, 2*(3), 39–65.

Chiucchi, M. S., & Giuliani, M. (2017). Who's on Stage? The roles of the project sponsor and of the project leader in ic reporting. *Electronic Journal of Knowledge Management, 15*(3), 183–193.

Chiucchi, M. S., Giuliani, M., & Marasca, S. (2014). The design, implementation and use of intellectual capital measurements: A case study. *Management Control, 2,* 143–168.

Chiucchi, M. S., Giuliani, M., & Marasca, S. (2016). The use of intellectual capital reports: The case of Italy. *Electronic Journal of Knowledge Management, 14*(4), 245–255.

Chiucchi, M. S., Giuliani, M., & Marasca, S. (2018a). Levers and barriers to the implementation of intellectual capital reports: A field study. In J. Guthrie, J. Dumay, F. Ricceri, & C. Nielsen (Eds.), *The Routledge companion to intellectual capital* (pp. 332–346). New York: Routledge.

Chiucchi, M. S., & Marasca, S. (2004). La progettazione del sistema di misurazione del capitale intellettuale: Il caso iGuzzini illuminazione. In AA.VV (Eds.), *Knowledge management e Successo Aziendale* (pp. 1095–1114). Udine: Arti Grafiche Friulane.

Chiucchi, M. S., & Montemari, M. (2016). Investigating the 'fate' of intellectual capital indicators: A case study. *Journal of Intellectual Capital, 17*(2), 238–254.

Chiucchi, M. S., Montemari, M., & Gatti, M. (2018b). The influence of integrated reporting on management control systems: A case study. *International Journal of Business and Management, 13*(7), 19–32.

D'Andrea, A. (2012). Verso la "sustainability accounting" in Sanità: Il caso della Medicina Trasfusionale. *Rivista di Studi Sulla Sostenibilità, 1,* 97–123.

D'Andrea, A. (2017). Applying GRI sustainability reporting in the water sector: Evidences from an italian company. *International Journal of Business Administration, 8*(3), 10–23.

De Santis, F., & Giuliani, M. (2014). Prisoners of inertia. Reflections on the auditors' everyday practice. In *Proceedings of the European Accounting Association Conference*, Tallinn, May 21–23, 2013.

De Santis, F., & Giuliani, M. (2015). You are what you do. Investigating the role of auditors in practice. In *Proceedings of the European Accounting Association Conference*, Glasgow, April 28–30, 2015.

Del Bene, L., & Ceccarelli, R. (2016). Il contributo del Sistema Informativo alla gestione logistica dei materiali. Un caso di studio nel settore sanitario. *Management Control, 1,* 149–172.

Fejér-Király, G. (2015). Bankruptcy prediction: a survey on evolution, critiques, and solutions. *Acta Universitatis Sapientiae, Economics and Business, 3*(1), 93–108.

Ferreira, A., & Otley, D. (2009). The design and use of performance management systems: An extended framework for analysis. *Management Accounting Research, 20*(4), 263–282.

Gatti, M. (2011a). *Balanced scorecard e cost management. Riferimenti teorici e casi aziendali.* Bologna: Esculapio Editore.

Gatti, M. (2011b). Gli strumenti di controllo orientati al cliente: Un'analisi sistemica. *Management Control, 1*(1), 99–124.

Gatti, M. (2013). Opening the "black box" of strategy: The role of management accounting systems. *International Journal of Finance and Accounting, 2*(7), 393–399.

Gatti, M. (2015). Exploring the challenges of measuring intangibles. The implementation of a Balanced Scorecard in an Italian company. *International Journal of Management Cases, 17*(4), 120–133.

Gatti, M., & Chiucchi, M. S. (2017). Context matters. Il ruolo del contesto negli studi di controllo di gestione. *Management Control, 3,* 5–10.

Gatti, M., Chiucchi, M. S., & Montemari, M. (2018). Management control systems and integrated reporting: Which relationships? The case of the Azienda Ospedaliero Universitaria Ospedali Riuniti Ancona. *International Journal of Business and Management, 13*(9), 169–181.

Gatti, M., & Poli, S. (2011). L'evoluzione del sistema contabile e di controllo della Santa Casa di Loreto tra il XVI e il XVIII Secolo: Un'interpretazione tra sacro e profano. *Contabilità e Cultura Aziendale, 11*(1), 5–30.

Gatti, M., & Poli, S. (2014). Accounting and the papal states: The influence of the pro commissa Bull (1592) on the rise of an early modern state. *Accounting History, 19*(4), 475–506.

Gatti, M., & Poli, S. (2018). Accounting and political parties: Explaining the 'why' of an Italian light touch regulation. *Accounting, Auditing & Accountability Journal, 31*(6), 1618–1643.

Giuliani, M. (2003). La valutazione del capitale intellettuale. Un approccio integrato. *Revisione Contabile, 51.*

Giuliani, M. (2006). La misurazione del capitale intellettuale: Utilità e limiti. *Revisione Contabile, 68,* 19–31.

Giuliani, M. (2010). Il controllo e la pubblicità del bilancio di gruppo. In S. Branciari, L. Marchi, & M. Zavani (Eds.), *Economia dei gruppi e bilancio consolidato* (3rd ed., pp. 231–244). Torino: Giappichelli.

Giuliani, M. (2014). Accounting for intellectual capital: Investigating reliability. *International Journal of Finance and Accounting, 3*(6), 341–348.

Giuliani, M. (2015). Intellectual capital dynamics: Seeing them "in practice" through a temporal lens. *VINE Journal of Information & Knowledge Management, 45*(1), 46–66.

Giuliani, M. (2016a). *La valutazione del capitale intellettuale*. Milano: Franco Angeli.

Giuliani, M. (2016b). Sensemaking, sensegiving and sense breaking: The case of intellectual capital measurements. *Journal of Intellectual Capital, 17*(2), 218–237.

Giuliani, M., & Brännström, D. (2011). Defining goodwill: A practice perspective. *Journal of Financial Reporting and Accounting, 9*(2), 161–175.

Giuliani M, Chiucchi MS (in press) Guess who's coming to dinner: The case of IC reporting in Italy. *Journal of Management and Governance*. https://doi.org/10.1007/s10997-018-9432-x.

Giuliani, M., Chiucchi, M. S., & Marasca, S. (2016). A history of intellectual capital measurements: From production to consumption. *Journal of Intellectual Capital, 17*(3), 590–606.

Giuliani, M., & Marasca, S. (2011). Construction and valuation of intellectual capital: a case study. *Journal of Intellectual Capital, 12*(3), 377–391.

Giuliani, M., & Marasca, S. (2017). La valutazione delle aziende bancarie. *Rivista dei Dottori Commercialisti, 2,* 217–241.

Giuliani, M., & Marasca, S. (Eds.). (2018a). *La valutazione degli intangibles aziendali*. Milano: Giuffrè.

Giuliani, M., & Marasca, S. (2018b). Gli approcci valutativi alla stima del marchio: Un'introduzione. *Rivista dei Dottori Commercialisti, 3,* 463–464.

Giuliani, M., & Skoog, M. (2017). Making sense of the temporal dimension of intellectual capital: A critical case study. *Critical Perspectives on Accounting*. https://doi.org/10.1016/j.cpa.2017.04.001.

Jönsson, S., & Lukka, K. (2007). There and back again: Doing interventionist research in management accounting. In C. S. Chapman, A. G. Hopwood, & M. D. Shields (Eds.), *Handbook of management accounting research* (pp. 373–397). Oxford: Elsevier.

Lambert, S., & Montemari, M. (2017). Business model research: From concepts to theories. *International Journal of Business and Management, 12*(11), 41–51.

Lüttgens, D., & Montemari, M. (2016). Editorial: New ways of developing and analyzing business model innovation. *Journal of Business Models, 4*(3), 1–4.

Marasca, S., & Chiucchi, M. S. (2008). La BSC come promotore del cambiamento del sistema di controllo di gestione: Il caso della Lega del Filo d'Oro. *Finanz Mark e Prod, 26*(1), 15–41.

Marasca, S., & Del Bene, L. (2009). Misurare le performance per migliorare la gestione: Il caso della Provincia di Ancona. *Azienda Pubblica, 2,* 285–310.

Marasca, S., & Giuliani, M. (2009). Il processo di valutazione degli intangibles: Riflessioni critiche su un caso aziendale. In E. Comuzzi, S. Marasca, & L. Olivotto (Eds.), *Intangibles. Profili di gestione e di misurazione* (pp. 183–204). Milano: Franco Angeli.

Marasca, S., Montanini, L., D'Andrea, A., & Cerioni, E. (2017). Adapting the integrated reporting framework to the public sector: A case study in healthcare. In *Proceedings of the 13th Interdisciplinary Workshop on Intangibles, Intellectual Capital and Extra-Financial Information*, Ancona, September 21–22, 2017.

Marasca, S., Montanini, L., Manelli, A., et al. (2018). Social reporting in a health care organization: A case study of a regional Italian hospital. In G. Gal, O. Akisik, & W. Wooldridge (Eds.), *Sustainability and social responsibility: Regulation and reporting. Accounting, finance, sustainability, governance & fraud: Theory and application* (pp. 333–367). Singapore: Springer.

Montanini, L., & D'Andrea, A. (2012). Medicina trasfusionale: Definizione e applicazione di un modello di accountability nella Regione Marche. *Mecosan, 83,* 87–102.

Montemari, M. (2018a). *Programmazione e controllo negli Atenei—Strumenti a supporto della governance in un contesto evolutivo.* Torino: Giappichelli.

Montemari, M. (2018b). Editorial: Introduction to the special issue based on papers presented at the business model conference 2018. *Journal of Business Models, 6*(2), 1–4.

Montemari, M., & Chiucchi, M. S. (2013). The problematization of IC indicators in action. In *Proceedings of the 9th EIASM Interdisciplinary Workshop on "Intangibles, Intellectual Capital & Extra-financial Information".* Copenhagen, Denmark, September 26–27, 2013.

Montemari, M., & Chiucchi, M. S. (2017). Enabling intellectual capital measurement through business model mapping: The Nexus case. In J. Guthrie, J. Dumay, F. Ricceri, & C. Nielsen (Eds.), *The Routledge companion to intellectual capital* (pp. 266–283). London: Routledge.

Montemari, M., & Nielsen, C. (2013). The role of causal maps in intellectual capital measurement and management. *Journal of Intellectual Capital, 14*(4), 522–546.

Montemari, M., & Nielsen, C. (2014). Value creation maps. In C. Nielsen & M. Lund (Eds.), *Business model design: Networking, innovating and globalizing* (pp. 13–25) Copenhagen: BookBoon.com/Ventus Publishing Aps.

Neely, A., & Bourne, M. (2000). Why measurement initiatives fail. *Measuring Business Excellence, 4*(4), 3–7.

Nespeca, A., & Chiucchi, M. S. (2018). The impact of business intelligence systems on management accounting systems: The consultant's perspective. In R. Lamboglia, A. Cardoni, R. Dameri, & D. Mancini (Eds.), *Network, smart and open* (pp. 283–297). Lecture notes in information systems and organisation Cham: Springer.

Nielsen, C., Lund, M., Montemari, M., et al. (2018a). *Business models: A research overview.* New York: Routledge.

Nielsen, C., Lund, M., Schaper, S., et al. (2018b). Depicting a performative research Agenda: The 4th stage of business model research. *Journal of Business Models, 6*(2), 59–64.

Nielsen, C., & Montemari, M. (2012). The role of human resources in business model performance: The case of network-based companies. *Journal of Human Resource Costing & Accounting, 16*(2), 142–164.

Poli, S. (2008). L'impairment test delle immobilizzazioni materiali secondo l'OIC 16: Analisi di alcune criticità. *Revisione Contabile, 81,* 18–33.

Poli, S. (2011). Bilanci di previsione delle comunità pontificie secondo la bolla Pro commissa (15 agosto 1592). *Rivista Italiana di Ragioneria e di Economia Aziendale, 111*(11–12), 678–689.

Poli, S. (2012a). *La svalutazione delle immobilizzazioni materiali nei bilanci delle imprese italiane.* Torino: Giappichelli.

Poli, S. (2012b). Accounting e "buon governo" delle comunità locali dello Stato Pontificio alla fine del XVI Secolo. *Contabilità e Cultura Aziendale, 12*(1), 156–178.

Poli, S. (2013a). The application of the accounting concept of materiality in the Italian listed companies' financial statements. *International Journal of Finance and Accounting, 2*(4), 214–219.

Poli, S. (2013b). Small-sized companies' earnings management: Evidence from Italy. *International Journal of Accounting and Financial Reporting, 3*(2), 93–109.

Poli, S. (2013c). The Italian unlisted companies' earnings management practices: The impacts of fiscal and financial incentives. *Research Journal of Finance and Accounting, 4*(11), 48–60.

Poli, S. (2015a). The links between accounting and tax reporting: the case of bad debt expense in the Italian context. *International Business Research, 8*(5), 93–100.

Poli, S. (2015b). Do ownership structure characteristics affect Italian private companies' propensity to engage in the practices of "earnings minimization" and "earnings change minimization"? *International Journal of Economics and Finance, 7*(6), 193–207.

Poli, S. (2017a). Is gender diversity in ownership structure related to private Italian companies' propensity to engage in earnings management practices? *African Journal of Business Management, 11*(1), 1–11.

Poli, S. (2017b). The relationship between shareholder gender and earnings management in private Italian companies. *International Journal of Business, Management and Social Research, 12*(1), 11–27.

Roslender, R., & Fincham, R. (2004). Intellectual capital accounting in the UK: A field study perspective. *Accounting, Auditing & Accountability Journal, 17*(2), 178–209.

Taran, Y., Nielsen, C., Montemari, M., et al. (2016). Business model configurations: A five-V framework to map out potential innovation routes. *European Journal of Innovation Management, 19*(4), 492–527.

Finance, Law and New Technologies

Monica De Angelis, Filippo Fiordiponti, Erika Giorgini, Caterina Lucarelli, Camilla Mazzoli and Andrea Perna

Abstract Finance and technology—Fintech, from the *crasis* of Finance and Technology—are currently present in many everyday activities. Innovation is creating huge opportunities but also new threats, which deserve to be examined and monitored. As already stated by the Financial Stability Board, a global approach is essential, however it needs to be shared with emerging economies. Technology and "Big Data" are changing the shapes of financial industry. Such a revolution pushes operators, investors and regulators towards unexplored territories, so that the legal and economic questions deriving therefrom need to be answered. The Business Science Department has started an interdisciplinary study, whose interesting results will be verified and delved into in the next two years.

M. De Angelis (✉) · F. Fiordiponti · E. Giorgini · C. Lucarelli · C. Mazzoli · A. Perna
Department of Management, Università Politecnica delle Marche, Ancona, Italy
e-mail: m.deangelis@univpm.it

F. Fiordiponti
e-mail: f.fiordiponti@univpm.it

E. Giorgini
e-mail: erika.giorgini@univpm.it

C. Lucarelli
e-mail: c.lucarelli@univpm.it

C. Mazzoli
e-mail: c.mazzoli@univpm.it

A. Perna
e-mail: a.perna@univpm.it

© Springer Nature Switzerland AG 2019
S. Longhi et al. (eds.), *The First Outstanding 50 Years
of "Università Politecnica delle Marche"*,
https://doi.org/10.1007/978-3-030-33879-4_13

1   Introduction

Nowadays Finance and Technology—Fintech from the crasis between the words Finance and Technology—are present in many daily activities.

From the simplest gesture of sending a payment with mobile phone, to using virtual currencies, from collecting online funds (crowdfunding), to direct internet loans (peer to peer lending), up to the robo-advice for financial advisory services to customers.

The possibility of offering new products or renewing the methods of providing financial services has widened the audience of users who are no longer restricted to the "experts" of the banking and financial world. In this context, regulators—from the International Monetary Fund to the European Commission, to our Bank of Italy and Consob—must be animated by two deep awareness: law follows innovation, it cannot anticipate it; innovation opens up immense opportunities which, however, entail new risks that must be studied and monitored.

It was authoritatively (Christine Lagarde) reiterated the need for a common approach to Fintech that takes into account the three lessons of the great financial crisis. First: trust is the foundation of the financial system, but that it can easily be shaken. It is therefore necessary to grasp the benefits of technological innovations to produce new and stronger confidence. Second: the years before the great crisis have seen the proliferation of financial instruments—such as bonds secured by debts—whose risk was not fully understood by investors.

It is therefore crucial to understand whether, with a vast wealth of information, how much decentralized thanks to Fintech, risks will multiply rather, be reduced, if they are more fragmented and difficult to identify or if there are simply new and different risks. It will also be investigated whether the increasingly weakened role of traditional intermediaries will also have repercussions on the adequacy of the information set offered to the investor. Third: in a globalized world, financial shocks spread beyond national borders and the response to crises requires, therefore, more and more common action.

If, as mentioned, innovation is continuous and the right can only follow it, the Authorities have the responsibility of developing a regulatory system capable of looking ahead, with flexibility and new knowledge.

This is why a global approach is essential, which has already been launched by the Financial Stability Board, but which must be extended and shared with emerging economies. Ultimately, technology and "big data" are changing the connotations of the financial industry. It is a revolution that pushes operators, investors, regulators on unexplored terrain and requires answers to the numerous legal and economic questions that derive from it.

Before going into the specific world of financial innovation, some general considerations on the relationship between law and technology seem useful.

1.1 Some General Notes on the Relationship Between Rules and Technologies

Technologies, digitalization and innovation are a challenge for every legislator world-wide (De Angelis 2017): the wide use of robotics, for instance, shapes the finance sector faster than the rules could give an adequate response. Moreover, distinctive elements of the current technological era challenge traditional legal categories and raise new regulatory concerns.

In an age where technology and innovation are pillars of society, analyzing the juridical issues (such as constitutional principles, intersecting rules, etc.) arise when scientific discoveries are inserted in the financial system is ever more complex: this because the scientific progress, nourished by new technologies (this is the reason why is it possible to talk about technological progress), requires answers often difficult to be identified. Primarily because the legal dimension of the issues produced by technologies sets off many consequences in a number of areas: just think of privacy. Furthermore, either the answers generated by laws, either application practices should in addition reflect the different orientations proposed and showed by various legal cultures.

The considerations related in this theme are many, in the following lines there will be a focus only some of them. The principal aim of this part of the article is to underline that technology and science (in general) must be regulated but if there is a need for regulation, it is necessary for this regulation to operate according to the specific characteristic of the sector, of the service, of the context, of the particularities of the right under review. We also need to consider that the introduction of rules regarding scientific progress is not simple due to the transnationality. So, there will be situations which a stricter frame of regulation will be needed for and other cases which slightly hinted frame will be needed for.

Even if the codification of the new phenomena doesn't encounter particular juridical obstacles, the scientific progress and technological development need attention from at least two corners: new drafting methodologies by traditional standards (these new methodologies provide for different paths for the construction of upstream and downstream rules and new ways of training the rule makers); and the development resilient organizations within the financial system that is able to resist, to absorb, to adapt and recover in an efficient and timely manner from the effects of new regulations. This may need the ongoing protection and recovery of its structures and essential functions.

Therefore, new technologies require a new connection between rules and financial system and the parties who apply those rules: for example, technologies impose a different way of thinking about the protection of consumers rights' targeted at their specific needs; technologies affect how traditional legal principles and arrangements are applied: for example, proportionality, privacy, responsibility and justiciability. Moreover, company structures and planning services must be modulated and adapted to reconcile new innovations and established practice.

Basically, these lines want to underline that is not simply a matter of creating and imposing rules on the financial systems run over by the scientific progress. In these instances it would appear to be more convenient to apply perspective principles rather than static rules and have a comprehensive vision of the financial sector in the hands of the actors who will implement those principles (so called anticipatory governance).

In an evolving system as financial is, it is necessary to face a high complexity that comes not only from the intrinsic characteristics of the different actors involved (customers, providers, professional groups, companies, organizations) but also by needs posed by institutional and technical sub-system.

All these actors will have to confront the legislative "hetero imposition" and at the same time they have to demonstrate ability to be resilient and manage the rules through adaptive modalities (so called management of rules).

Technology itself (we refer in particular to ICT) helps in the pursuit of this intent: the systems are increasingly digitalized and processes and decisions also follow this rhythm. In the last years communication or contact among actors (public and privates ones) is increasingly done via technologies. In this way technologies make the relationship among actors easier and simpler.

In fact, technology enhances communication, having a multiplier effect on relationships: e.g. the intelligent use of continuous stream of data and information enables customers to interact with the services' providers in order to improve performance.

The players of the financial system should move beyond a traditional use of digital technologies (that is doing quickly what previously was done in analogic), because the aim is not to create a "digital customer of a smart financial system", but to use technology (and not only ICT) to reorganize services to ensure its use according to strategies relating to satisfaction.

Technologies applied to the financial sector (and not only to finance), should be useful in the care of customers' interests, and should be closely related to scientific progress as well as to the improvement of the financial system as a whole. This relationship should be obvious as technologies impose technical issues. These issues rise, in fact, arise even in political arena: for instance, the choice to legislate or not legislate on topics such as the neutrality of Internet, or the responsibility of the professionals in the use of cybernetics. In other words, when the legislator can choose whether to legislate or not, he's making a political choice, not a technical one. Moreover, when the rule maker chooses to intervene it is assumed that it is in order to protect public interests: the choice involves not only the an (if we should introduce rules) but also the quid (what rules) and quomodo (in which way we can regulate).

It is not easy to determine the way to legislate and when legislator should intervene on situations invested by new technologies. In any case, the legislator must take into account the political context where to act, if the purpose is to have an efficient and effective result.

Preferably technological innovation should have rules after its effects have been unfolded: this is because if we want useful rules, we have to observe what kind of consequences are being produced by the technologies.

The introduction of mandatory rules should be applied at a later stage; this is because binding rules may cause more difficulties with the innovations. In order to guide and influence, the choice to set some rules before the complete effects are known, should be made as general guidelines, definitions and details should be left to a later stage, keeping in mind that any rule should respect fundamental rights and have less adverse effect as possible on the legal guarantees of interested parties.

Otherwise, despite the usefulness of technology, the decision to rule it may not be helpful: hence the opportunity to consider a soft law approach (we ask for not only useful technologies but also useful rules on technology). This approach would appear to be better because it better respects the specificity of the financial system in an era whose response time are different from the traditional ones.

Finally, do not forget that we can demand elasticity to regulations in order to maximize adaptation the rapidity of scientific progress and—at the same time—provide reasonable protection to the rights of all actors involved.

1.2 Financial Advisory Services in the Fintech Era

In this era of digitalisation the whole world has formatted in an extension compatible with every computer: numbers.[1] Consequently, a large amount of data have been produced which, since then, could have been transformed and elaborated in other data ready to circulate thanks to fundamental tools and, nonetheless, substantially different such as: the Internet and computers.

This new phenomenon, in the financial field, has been called, as already said, Fintech. Even today, however, there is no universally recognized definition that can summarize the variegated ways in which this phenomenon is expressed, both in the manifestations of the present and, even more, in the future. A correct juridical characterization of this phenomenon passes first of all from the comprehension of the characteristics and modes of operation of the means or technologies used. Only in this way can it be ascertained whether a given service or financial activity, for the new ways in which it is expressed, requires some form of intervention in the regulatory system that regulates it. It is not by chance that technology, or rather the methods of its application, are relevant to the service or activity to which it is accessed. Consider, for example, the service of Fintech that, due to the heterogeneity of the forms of interaction with the technological means, offers the major points of reflection such as robo-advice.

Considering the objectives of this work it becomes important to start from the notion of advice and its related discipline.

The primary regulation defines the "advice on investments" as provision of personal recommendations to a client, either upon his request or at the initiative of the service provider, with respect to one or more transactions relating to a particular

---

[1] Esposito (2017) affirms that «the premise is the process of "datification", which allows us to express more and more phenomena in a quantified format that can be analyzed and processed».

financial instrument. The recommendation is to be understood as being customized when presented as suitable for the client or is based on a consideration of the customer's characteristics. In contrast, a recommendation is not personalized if it is circulated to the public through distribution channels.

With this in mind, the advice would represent all the times that the recommendation is represented to the customer, or at least to appear to the client, as the result of comparison with their own individual characteristics, as well as those times that involve not only a financial instrument but where it relates to various types of financial instruments, including a financial planning (Giorgini 2017).

There is no doubt that technology currently plays a key role in this type of financial service, so much so that the term automated consultancy is now in use.

There are at least three types of automated advice or robo-advice. Depending on the level of automation and its allocation in the service execution process, the following is identified: (a) a pure robo-advice in which automation characterizes all phases of the service use process, from initial customer profiling to construction of the investment portfolio subject to the recommendation; (b) a hybrid robo-advice that combines and/or alternates the human element and the digital element in one or more phases of the value chain. In both cases, the service is directed to the final investor; (c) a third model known as robo-advisor that puts the automated tools in support of the consultant, thus qualifying as B2B.

But there is more. The automation combined with the telematic infrastructure, thanks also to the new big data management and analysis capabilities, could in fact give rise to distortive scenarios. It could happen for example, that only one robo-advisor or, alternatively, multiple networks of financial consultants supported by a single robo-advisor, assist a potentially endless number of customers.

Information or personal data is of primary importance in order to allow the robo-advisor the level of personalization required by the discipline enacted by MiFIDII. The principle of the Know your customer rule, on which the canons of adequacy and appropriateness in fact imposes on the financial consultant a greater and more penetrating analysis of the client's financial knowledge, his financial position, the propensity to risk, the level of education, etc. ...

But this data (Big Data),[2] coming from different sources such as web pages, files from weblog, forum, social-media, audios, videos, click streams, emails, documents and sensor systems automatically implement new data, constituting new assets (Big Data Economy) (Perlingieri 2018). Such new assets, from a legal point of view represent «new situations, assets mostly in the users' availability» (Perlingieri 2017).

In digitalism era, in fact, what is radically different is the volume, variability and speed of data, as well as the possibility to link them one with the other by algorithms (Giorgini 2019).

---

[2] «With the expression "big data" it is refereed to large amounts of data of different types, produced at high speed, by different types of sources. The management of these dataset at high variability and at real time imposes to resort to new instruments and methodologies, such as for example, powerful processors, software, and algorithms»: declaration Communication by the Commission, Verso una florida economia basata sui dati (COM(2014)0442).

It must be stressed that "the code" created by the algorithms is for its own nature designated to produce approximated results in a double direction. On one hand, the digital binary numerical system requires to constantly approximate the data at each operation. On the other, the simplification and reduction aimed at making efficient the algorithm produces errors which are directly proportional to the increasing of the number size. All that is aggravated by the fact that the multiple operations of calculation are hidden and, therefore, out of control of the human user[3] and that can also create serious discriminatory phenomena.

The scenario described, even if briefly, is further complicated if only attention is focused on the growing use of the deep learning methodology in which is not possible to know ex ante the data that are the inputs of the algorithmic procedure. Thus, the data are not provided by humans but they are learned from the algorithm itself (Giorgini 2019).

So, it could then happen that the automation and subsequent massification of the financial advisory service cannot guarantee effective personalization of the service. A useful example is the "mass market" whose business model provides a standardized consulting service and a range of products offered reduced due to the availability of customers in terms of fees and the consequent lower returns from the robo-advisor. In this hypothesis there is the risk that the algorithm, for efficiency reasons, prefers to operate in simplified contexts and for the effect to define a closed number of classes of investors to assign, according to the risk profile, the same investment recommendation.

In conclusion, the vision—pursued by the legislator of the financial markets,[4] of monitoring the results of the algorithmic application that would subsequently allow to intervene on the construction of the algorithm itself—is not sufficient.

The functioning of algorithms involving legal effects must be controlled to ensure that they produce outcomes compatible with our legal system. This control must take place through the interpretation of the act generated by the application of the algorithm. It cannot be sufficient, even if indispensable, to rely on the right to access to the code, or to implement measures to safeguard (Giorgini 2019).

2 Innovations in Financial Systems

The financial industry has always been an early adopter of new technologies: starting from the '60, the need of clients to get cash outside normal working hours has led to the introduction of Automatic Teller Machines; since then, many automated services such as debit and credit cards, online trading and online banking services have been

---

[3]C. O'Neil, Weapons of math destruction: how big data increases inequality and threatens democracy, New York, 2016, p. 19; L. Avitabile, Il diritto davanti all'algoritmo, in Riv. it. per le scienze giuridiche, 2017, p. 321. In the same meaning, see: Guidelines on Automated individual decision-making and Profiling for the purposes of Regulation 2016/679, «Profiling processes can be opaque. Individuals might not know that they are being profiled or understand what is involved».
[4]See: artt. 4(1), n. 39 e 17 MiFID II.

offered in response to customers' demand for immediacy. Therefore, the widespread adoption of digital technology in financial markets in recent years should not come as a surprise (Panetta 2018).

Nevertheless, differently from the past, technology is now invading almost each and every single aspect of business in the financial industry, thus giving rise to a new era which has been recently labeled as 'Fintech'. Fintech represents 'technologically enabled financial innovation that could result in new business models, applications, processes or products with an associated material effect on financial markets and institutions and the provision of financial services' (Financial Stability Board 2017a). In other words, Fintech refers to new digital technologies which automate a wide range of financial activities and may provide new and more cost-effective products in the financial sector, ranging from lending to asset management, and from port-folio advice to the payment systems. Despite Fintech's business volume is small in comparison to the size of financially intermediated assets and capital markets, it has the potential to disrupt established financial intermediaries (Vives 2017). In this new context, the financial sector—and banks in particular—are forced to bring about a deep transformation but also regulators and supervisors are now standing new and complex challenges in the next future (Carbò-Valverde 2017). The widespread adoption of digital technology in financial markets is reshaping both the landscape in the banking industry and also the lending relationship between firms and banks, moving from a bank-based to a market-base finance.

2.1 Fintech and Banking Systems

As far as the banking industry landscape is concerned, the last decades have witnessed a substantial number of mergers and acquisitions in the banking sector of many industrial countries, as documented in various official reports (see e.g. European Commission 2017; Financial Stability Board 2017; Bank for International Settlements 2018; Consob 2018) and research papers (Lozano-Vivas 2011; Schoenmaker 2015; Kowalik et al. 2015). Concentration ratios—measured as the share of banking system assets held by the largest five banks—have recently increased within the euro area and the United States (Bank for International Settlements 2018); despite such consolidation is partly a consequence of dealing with the effects of the crisis, the increasing concentration that has been registered in some countries that were less affected by the crisis suggests that such consolidation should better be explained by a number of concurrent factors: the communication technology progress, the deregulation, the general globalization and the resulting competitive challenges for financial firms. The technological progress, in particular, is a key driver of competition for firms and competition is often faced with consolidation. The Fintech revolution is no exception: a new wave of partnerships and acquisitions is expected in the next years involving incumbent banks and Fintech firms, as a result of the technological innovation that is overwhelming banks. Jeopardized by a multitude of Fintech start-ups, banks will need to consolidate in order to find room to invest in digital and, even more, banks must open up to the idea of working with Fintech firms, adding that it could be a win–win strategy for both sides.

Nonetheless, the strategies for new entrants and those of existing banks will depend on whether investment makes a firm tough or soft in the competition and on whether competition in the market place involves strategic substitutes or complements. Thus, depending on the underlying industry characteristics an incumbent bank may decide to accommodate or prevent entry (Vives 2017). Some authors (Barba Navaretti et al. 2017; Goodbody 2017) maintain that collaborations with Fintech firms represent an opportunity for banks to increase their efficiency both in the production and distribution of products and services, by means of an innovation, and subsequent improvement, of their supply. For example, in the presence of significant switching costs, an established incumbent bank is likely to maintain the status quo because it wants to protect the profitability of its large customer base. This may allow an entrant to enter and attract, for example, technology-oriented customers or even unbanked consumers. In these cases, Fintech firms have the potential to complement the retail banking services in a win–win strategy: if it not so, incumbent banks will have to work hard to create a true digital enterprise to compete with Fintech firms and, at the same time, while Fintech have developed applications that create improved customer experiences, they might lack skills in customer acquisition and other fields. That being so, collaborative partnerships might become increasingly important. Examples of such partnerships are already emerging either through joint partnerships, service outsourcing, venture capital funding, or acquisitions. In addition to this, banks may have to move quickly to identify acquisition targets before the most attractive FinTech firms are taken by competitors; the major risk for banks does not come from Fintech players but from banks that are better at partnering (Ernst and Young 2017).

Other authors support the idea that new Fintech competitors will harm incumbent banks because of banks' incapacity to innovate (Sperimborgo 2016; Bofondi 2017). According to such theory, incumbent banks may want to prevent or foreclose entry to new Fintech competitors. If it is so, new entrants could be possibly forced to rely on the payment infrastructure of the incumbent bank to offer complementary or differentiated services, in order to protect the established business of the bank.

In sum, in the next years incumbent banks will be forced to come to a decision on partnering with the new entrants, buying them up partially or totally, or fighting them. Indeed, the response that institutions will give is likely to be heterogeneous according to their specificity, to the extent of their legacy technologies and to the market conditions.

2.2   Fintech and Development of Market-Base Finance

As mentioned at the beginning of Sect. 2, the widespread adoption of digital technology in financial markets not only is reshaping the landscape in the banking industry, but it is also leading to a transfer from a bank-based to a market-base finance. Since

Fintech start-ups follow a customer-centric approach, new products and services (e.g. decentralized online peer-to-peer lending platforms) are being offered as opposed to the traditional bank-based products and services (Lin et al. 2015; Wang et al. 2015; Yan et al. 2015). New technology-driven solutions offer alternatives to the key banking services, especially with regard to the banks' two main functions: payments and lending.

As far as funding of loans is concerned, technology has no doubt facilitated peer-to-peer and marketplace funding. Digital technology allows direct finance with the matching of borrowers and investors. More specifically, Fintech uses innovative data analysis techniques (artificial intelligence and machine learning) to process efficiently the information that individuals and companies put online, sometimes unknowingly (big data). Using this data, algorithms calculate the creditworthiness of those applying for a loan and the result is available, on digital platforms not requiring a bank to act as intermediary, to the savers who directly disburse the financing. Fintech firms are then undoubtedly changing the relationship between clients and financial service providers. On the one hand, this poses a serious threat to traditional banks but, on the other hand, as lending is not obviously just about matching investors and borrowers, banks still hold their competitive advantage in the value chain. The value chain of banks includes bundled services which provide powerful economies of scope to banks: the control of risk after lending has taken place, the trading of claims if investors need to access liquidity, and the management of non-performing assets. But even more, it involves trust and the familiarity with other human beings behind a counter. Differently from banks, Fintech firms generally carry out one or more of the above-mentioned activities in an unbundled way so that, if Fintech firms wish to expand their businesses to exploit such economies of scope, then they will probably have to transform themselves into banks. Certainly, banks will have to give much more importance to the digital distribution channels and radically alter the way that client data are analyzed and stored. Substantial investment in technology and human capital will be necessary in order to protect their business.

More than once in the past authors questioned the role and the future of banks as a consequence of innovations in products and service. For example, when securitization was introduced the question was whether securitization would undermine the banks' lending role. Boyd and Gertler (1994) demonstrated that while securitization would make banks less important for the actual funding of loans, the core functions of banks in the lending process would be preserved, as would the centrality of banks.

The message of that article is undoubtedly relevant today (Darmine 2017). Banks need to respond to the new technological challenges and try to be players in the Fintech world themselves. Moreover, the distinctions by types of provider, function, or friction is becoming more and more blurry. The next future is one where many traditional classifications, like the dichotomy banks versus markets, are increasingly less relevant; the financial market landscape is in a state of flux and the final outcome in terms of the financial market structure and competition is difficult to anticipate.

3 Bringing Innovation to the Industry: The Role of Universities

On the company side, SMEs and start-ups have frequently claimed their difficulties in accessing to traditional bank loans, especially after the crisis. As anticipated in preceding Sections, new types of financing are emerging with Fintech development, also thanks to institutional and technological developments. However, the response of the national financial systems is still uneven across European countries, and a deeper analysis is needed to find the way in which the financial sector may become fully supportive of the development of new ventures.

Given this promising research area, we plan to investigate how the technology transfer process change within the paradigm of Industry 4.0, being supported by both the regulatory and the financial framework offered by Fintech.

In order to reach this final long-term research goal, a short-term forthcoming investigation refers the field of entrepreneurship and technology transfer that request the analysis of how new business ventures are created and developed by adopting an industrial network perspective: particularly, we are going to explore how the combination of resources, controlled by other actors, impact the new business formation process.

With reference to the new business creation, previous research of the team has investigated the development of new business relationships and early stages of business relationships with the purpose of understanding the role of the exogenous factors. The team have analyzed in-depth how external circumstances affect the focal relationship development. In fact, forming a new venture is a matter of engaging with multiple actors (Snehota 2011) since the focal company doesn't own all the necessary resources (money, people, technologies, etc.). According to the IMP perspective (Håkansson and Snehota 1995) understanding the formation and development of new ventures has implied to look at how key processes—such as research and development, technology development and finance—unfold and interlink with each other. Moreover, what IMP recognizes as fundamental is the central role played by interactions and business relationships to let the new venture to grow. Therefore, in contrast with entrepreneurship scholars IMP focuses on the relevance of processes and how key business processes link to the external context of the newly formed company (La Rocca and Perna 2014).

The current social and economic setting is characterized by an increasing number of new technologies (IoT, big data, etc.) which are supposed to speed up business processes, social processes etc.: what are, therefore, the consequences derived from the use of such technologies when new business is formed? Better to say, are the new upcoming firms aware about the challenges posed by Industry 4.0 and digitalization?

It is clear that new actors, new processes, new business models are emerging in order to cope with the issue of creating something new a much more fast and global and ever changing context, therefore we may think that it is not only a 'new industrial revolution', it is more a new way of 'interacting' and 'relating' to a different and perhaps more complex social-economic context.

3.1 The University-Industry Interaction

Turning to how new business is formed within an emergent new paradigm, an interesting and promising research avenue deals with: (1) an increasing understanding of the role played by the technology and digitalization as mechanism supposed to facilitate new business formation; (2) an increasing understanding of the role of organizations such as universities, research centers, incubators etc. as institutions which may supply to the nascent business important resources (knowledge, innovation, etc.) and help to perform specific activities (training, search of economic resources, business coaching etc.).

More precisely, our research group is going to shed more light on point (2) since it is an interesting topic within the stream of research which look at how University-Industry interact in order to develop new knowledge, implement specific technologies, etc. Moreover, although studying the phenomenon of new business formation as consequence of University-Industry collaborations would deserve more attention, it is still very much unexplored how university and industry collaborates and what are the outcomes of the collaborations: another central question would focus on which counterpart gains what over the collaboration. 'Commercializing science' (Baraldi and Ingemansson-Havenvid 2016; Baraldi and Waluszewski 2011) for instance would represent an interesting setting to look at in order to understand University-Industry dynamics.

Creating a new product and transforming it into a marketable solution within both academic and industrial settings is often a long process, time and resource consuming. In order to tackle the issue of commercialization there is an obvious need of zooming into the interactive processes which characterize several different actors such as universities, private companies, third parties, etc.

3.2 The Empirical Analysis: Eureka and Italy-Sweden Comparison

Consider that our focus, here, is the study of University-Industry interactions organized in order to develop new knowledge and implement specific technologies, thus understanding pros and cons of the Universities-enterprises relationship.

For this purpose, we are running an empirical research in the Marche region in Italy based on the case of EUREKA project. The EUREKA project resembles an enforcement of concept of the Triple Helix of university-industry-government relationships initiated in the 1990s by Etzkowitz (1998) and Etzkowitz and Leydesdorff (2000). It is based on a co-financing of PhD fellowships with a balanced involvement of three actors: (i) the government of Marche Region; (ii) the four Universities of this Region (Università Politecnica Marche, Università di Urbino, Università di Camerino and Università di Macerata); (iii) and a selection of SMEs of the same regional territory, on the Industry side.

The strength of our research design relies on the fact that exploration of the EUREKA projects is managed within a cross-country comparison. Precisely, we compare Italy, with the case of Marche Region, with Sweden which is ranked first in the European rankings in terms of innovation aptitude (Sweden in first place of the European Innovation Scoreboard 2017). The interest and usefulness of the research is confirmed by the active involvement of Uppsala University Innovation centre[5] (UU Innovation) which is partly owned by Uppsala University whose aim is to spread collaborative innovations within the Uppsala-Stockholm region; UU innovation is administering the survey questionnaires and collecting data also from other two Swedish Universities very much connected with UU (University of Linköping and Örebro University). The operational phases are two: in a first phase, we will perform a survey on all Eureka projects, for all the four Universities of Marche, in symmetry with the three Swedish Universities; in a second phase, we will carry out a targeted study (qualitative analysis) on a selection of case studies (10 for Italy, 10 for Sweden). Each case study includes an interview protocol, both on the company side and on the university side (i.e., 40 interviews).

Both the survey questionnaires, for phase 1, and the phase 2 interviews guidelines, have been studied to be administered in the respective native languages, with a careful control of the interpretative correspondence passing through the 'bridge language', which obviously its English. Now, at the moment of the presentation of this paper, the first phase is underway. We completed the expedition of the survey to the tutors of all four University of Marche. This cleaning procedure[6] allowed us to identify the definitive sample, which consists of 129 EUREKA projects for Università Politecnica Marche, 87 for the Camerino University, 58 for Università di Macerata and 29 for Università di Urbino. At the beginning of 2019 we expect to submit the survey to the tutors of companies and to run the second phase of the research, with interviews to be held, simultaneously in the two countries.

We expected that this cross-country comparison, if and when it will be admissible, is expected to offer important insights, also in the direction of research policy.

---

[5]https://www.uuinnovation.uu.se/establish-collaboration/improving-methods/.

[6]Starting with the initial databases we received from the respective University Offices, we decided to undertake an initial check on the correctness of the data, by individually calling all the companies involved. For each of the Eureka partner companies we asked for confirmation of the name reported as TUTOR–Company side.

From the phone calls, we found critical issues, due to the unavailability of the tutor (in many cases leaked from the company) or even the company itself (due to bankruptcy or transformation/merger). Therefore, we adopted a policy to keep the case of the Eureka project in the observation sample if, to date, the company counterparty is still existing, or if the Company itself authorizes us to interview those who have followed the report (even if not formally Tutors) and it is still established in the company itself. Thanks to this control, we realized that there are two types of university tutors, to which only slightly different questionnaires were sent (for all four Universities): (a) those who have followed only one project (in the strict sense or made unique by the cleaning procedure mentioned above); (b) those who have followed more projects.

From this, we get that the number of people to whom we sent the survey does not coincide with the number of projects.

References

Ankrah, S., & Tabbaa, O. (2015). Universities–industry collaboration: A systematic review. *Scandinavian Journal of Management, 31*(3), 387–408.

Arner, D. W., Barberis, J. N., & Buckley, R. P. (2015). *The evolution of Fintech: A new post-crisis paradigm?* University of Hong Kong faculty of Law Research Paper 2015/047.

Avitabile, L. (2017). Il diritto davanti all'algoritmo. In *Rivista italiana per le scienze giuridiche* (p. 321).

Bank for International Settlements. (2018, February). *Sound practices: Implications of Fintech developments for banks and bank supervisors.*

Baraldi, E., & Ingemansson-Havenvid, M. (2016). Identifying new dimensions of business incubation: A multi-level analysis of Karolinska Institute's incubation system. *Technovation, 50–51,* 53–68.

Baraldi, E., & Waluszewski, A. (2011). Betting on science or muddling through the network two universities and one innovation commission. *The IMP Journal, 5*(2), 1–21.

Barba Navaretti, G., Calzolari, G., & Pozzolo, A. F. (2017). Fintech and banks: Friends or foes? *European Economy, 2,* 9–31.

Barnes, T., Pashby, I., & Gibbons, A. (2002). Effective university–industry interaction. *European Management Journal, 20*(3), 272–285.

Barringer, B. R. (2000). Walking a tightrope: Creating value through interorganizational relationships. *Journal of Management, 26*(3), 367–403.

Bertilsson Forsberg, P. (2018). *Collaboration in practice: A multiple case study on collaboration between small enterprises and university researchers.* Acta Universitatis Upsaliensis.

Bin, R., & Busatta, L. (2014). Introduzione. Forum: Law and the life sciences. In *Rivista di BioDiritto* (p. 7).

Bjerregaard, T. (2010). Industry and academia in convergence: Micro-institutional dimensions of R&D collaboration. *Technovation, 30*(2), 100–108.

Bofondi, M. (2017). Il lending-based crowdfunding: opportunità e rischi. In *Banca d'Italia, Questioni di Economia e Finanza* (No. 375).

Bonaccorsi, A., & Piccaluga, A. (1994). A theoretical framework for the evaluation of university-industry relationships. *R&D Management, 24*(3), 229–247.

Boyd, J. H., & Gertler, M. (1994). Are banks dead? Or are the reports greatly exaggerated? *Federal Reserve Bank of Minneapolis Quarterly Review, 18*(3), 2–23.

Bstieler, L., Hemmert, M., & Barczak, G. (2015). Trust formation in university-industry collaborations in the U.S. biotechnology industry: IP policies, shared governance, and champions. *Journal of Product Innovation Management, 32*(1), 111–121.

Carbò-Valverde, S. (2017). The impact on digitalization on banking and financial stability. *The Journal of Financial Management Markets and Institutions, 5*(1), 133–140.

Consob (Schena, C., Tanda, A., Arlotta, C., & Potenza, G.). (2018). Lo sviluppo del Fintech Opportunità e rischi per l'industria finanziaria nell'era digitale. *Quaderni Fintech* (Vol. 1).

Cyert, R. M., & Goodman, P. S. (1997). Creating effective university-industry alliances: An organizational learning perspective. *Organizational Dynamics, 25*(4), 45–57.

D'Este, P., & Patel, P. (2007). University–industry linkages in the UK: What are the factors underlying the variety of interactions with industry? *Research Policy, 36*(9), 1295–1313.

Darmine, J. (2017). Digital disruption and bank lending. *European Economy, 2,* 63–76.

De Angelis, M. (2017). Technologies and rules facing scientific progress in the healthcare system. In M. Conti & S. Orciani (Eds.), *Proceedings of 2016 International Workshop on Analysis of Biometric Parameters to Detect Relationships Between Stress and Sleep Quality (AnBiPa 2016).*

Duit, A. (2015). Resilience thinking: Lessons for public administration. *Public Administration, 94*(2).

Ernst & Young. (2017). *Scaling new heights. M&A integration in banking & capital markets.* Ernst & Young Global Limited.

Esposito, E. (2017). Artificial communication? The production of contingency by algorithms. *Zeitschrift für Soziologie*, 252.

Etzkowitz, H. (1998). The norms of entrepreneurial science: Cognitive effects of the new university–industry linkages. *Research Policy, 27*(8), 823–833.

Etzkowitz, H., & Leydesdorff, L. (2000). The dynamics of innovation: From National Systems and "Mode 2" to a Triple Helix of university–industry–government relations. *Research Policy, 29*(2), 109–123.

European Commission. (2017). Fintech: A more competitive and innovative European financial sector. Documento di consultazione.

Financial Stability Board. (2017, November). *Artificial intelligence and machine learning in financial services. Market developments and financial stability implications.*

Financial Stability Board (2017a). Financial Stability Implications from FinTech, Supervisory and Regulatory Issues that Merit Authorities' Attention, June 2017.

Giannaccari, A. (2017). La storia dei big data tra riflessioni teoriche e primi casi applicativi. In *Mercato concorrenza e regole* (p. 307).

Giorgini, E. (2017). *Consulenza finanziaria e sua adeguatezza, Napoli.*

Giorgini, E. (2019). Algorithms and law. *Italian Law Journal* (in press).

Goodbody. (2017). *UK banks – The challenger playbook.* Goodbody Stockbrokers UC.

Guffanti, E. (2011). Il servizio di consulenza: I confine della fattispecie. In *Società* (p. 556).

Håkansson, H., & Snehota, I. (1995). *Developing relationships in business networks.* London: Routledge.

Jonsson, L. (2015). Targeting academic engagement in open innovation: Tools, effects and challenges for university management. *Journal of the Knowledge Economy, 6*(3), 522–550.

Kowalik, M., Davig, T., Morris, C. S., & Regehr, K. (2015). Bank consolidation and merger activity following the crisis. *Economic Review, 1*, 31–49.

La Rocca, A., & Perna, A. (2014). New venture acquiring position in an existing network. *The IMP Journal, 8*(2), 28–37.

Lin, Z., Whinston, A., & Fan, S. (2015). Harnessing Internet finance with innovative cyber credit management. *Financial Innovation, 1*(5).

Linciano, N. (2012). La consulenza finanziaria tra errori di comportamento e conflitti di interesse. In *Analisi giuridica dell'economia* (p. 135).

López Martínez, R. E. (1994). Motivations and obstacles to university industry cooperation (UIC): A Mexican case. *R&D Management, 24*(1), 017–030.

Lozano-Vivas, A. (2011). Consolidation in the European banking industry: How effective is it? *Journal of Productivity Analysis, 36*(3), 247–261.

Moro Visconti, R. (2017). La valutazione economica dei database (banche dati). In *Il diritto industriale* (p. 358).

O'Neil, C. (2016). *Weapons of math destruction: How big data increases inequality and threatens democracy* (p. 19). New York.

Panetta, F. (2018). Fintech and Banking, today and tomorrow. Speech of the Deputy Governor of the Bank of Italy, Rome, 12th May 2018.

Paracampo, M. T. (2017). La consulenza finanziaria automatizzata. In *Fintech. Introduzione ai profili giuridici di un mercato unico tecnologico dei servizi finanziari, Torino* (p. 105).

Parrella, F. (2010). Consulenza in materia di investimenti. In *L'attuazione della MiFID in Italia, Bologna* (p. 183).

Parrella, F. (2011). Il contratto di consulenza finanziaria. In *Contratti del mercato finanziario, Torino.*

Peças, P., & Henriques, E. (2006). Best practices of collaboration between university and industrial SMEs. In A. Gunasekaran (Ed.), *Benchmarking: An International Journal, 13*(1/2), 54–67.

Perkmann, M., King, Z., & Pavelin, S. (2011). Engaging excellence? Effects of faculty quality on university engagement with industry. *Research Policy, 40*(4), 539–552.

Perlingieri, P. (2017). Relazione conclusiva. In C. Perlingieri & L. Ruggeri (Eds.), *Internet e diritto civile, Napoli* (p. 419).

Perlingieri, P. (2018). Privacy digitale e protezione dei dati personali tra per-sona e mercato. In *Foro napoletano* (p. 481).

Phan, P., Siegel, D. S. (2006). *The effectiveness of university technology transfer*.

Salazar, C. (2014). Umano, troppo umano…o no? Robot, androidi e cyborg nel "mondo del diritto" (prime notazioni). *Rivista di BioDiritto, 1*, 257.

Santoro, M. D., & Chakrabarti, A. K. (2001). Corporate strategic objectives for establishing relationships with university research centers. *IEEE Transactions on Engineering Management, 48*(2), 157–163.

Santoro, M. D., & Saparito, P. A. (2003). The firm's trust in its university partner as a key mediator in advancing knowledge and new technologies. *IEEE Transactions on Engineering Management, 50*(3), 362–373.

Schoenmaker, D. (2015). The new banking union landscape in Europe: Consolidation ahead? *Journal of Financial Perspectives, 3*(2), 189–201.

Sciarrone Alibrandi, A. (2009). Il servizio di consulenza in materia di investimenti: Profili di novità della fattispecie. In *L'attuazione della direttiva MiFID* (p. 87).

Siclari, D. (2016). La consulenza finanziaria indipendente prevista dalla Mifid II alla prova dei fatti. In *La MiFID II, Padova*.

Sironi, P. (2016). *Fintech innovation. From robo-advisor to goal based investing and gamification, Chichester* (p. 138).

Snehota, I. (2011). New business formation in business networks. *IMP Journal, 5*(1), 1–9.

Sperimborgo, S. (2016). Banche e innovazione tecnologica. Come avere successo nella tempesta perfetta della rivoluzione digitale. In Bancaria, No. 12.

Stella Ricther, M. (2016). Dalle mobili alle nobili frontiere della consulenza finanziaria. In *Banca borsa e titoli di credito* (Vol. I, p. 320).

Vives, X. (2017). The impact of Fintech on banking. *European Economy, 2*, 97–105.

Wang, H., Chen, K., Zhu, W., & Song, Z. (2015). A process model on P2P lending. *Financial Innovation, 1*(3).

Yan, J., Yu, W., & Zhao, J. (2015). How signaling and search costs affect information asymmetry in P2P lending: The economics of big data. *Financial Innovation, 1*(19).

Digital Transformation in B2B SMEs

Federica Pascucci, Valerio Temperini, Luca Marinelli
and Maria Rosaria Marcone

Abstract Nowadays digital transformation has a large impact on economy and business; the use of digital technologies can represent an important driver for companies in traditional and new markets. Many authors noticed that "digital is not an option" and the majority of the companies need to face this challenge. However different studies afford the topic for large enterprise in business to consumer context but at the same time, digital transformation is a relevant issue also for SME's in Business to Business context. SMEs can benefit from digitalization in facing the intense global competition, finding market opportunities, developing profitable relationships or transforming their business models. In this area many topics emerge from the literature and some of these seem to be particularly relevant: how social media could represent fruitful instruments for marketing and sales in b2b context; how digitalization can represent a sensible shortcut for the development of small businesses in foreign markets; how the use of e-learning tool could be a new way for marketing in SME's. The aim of the paper is to describe the role of digital transformation in SME's in business marketing and analyze challenges and opportunities for marketer and researchers.

F. Pascucci · V. Temperini · L. Marinelli (✉) · M. R. Marcone
Dipartimento di Management, Università Politecnica delle Marche, Ancona, Italy
e-mail: l.marinelli@univpm.it

F. Pascucci
e-mail: f.pascucci@univpm.it

V. Temperini
e-mail: v.temperini@univpm.it

M. R. Marcone
e-mail: m.r.marcone@univpm.it

© Springer Nature Switzerland AG 2019
S. Longhi et al. (eds.), *The First Outstanding 50 Years
of "Università Politecnica delle Marche"*,
https://doi.org/10.1007/978-3-030-33879-4_14

1   Introduction

The role of digital technologies in businesses is one of the most debated issues both in the academic field and among practitioners and policymakers. In particular, the focus is on how digital tools can effectively contribute to the creation of market opportunities, to the improvement of business management and ultimately to the firms' competitiveness. The transformation of digital technologies potential, in reality, is an interesting topic in particular with reference to the SMEs, for which the solutions offered by digitalization appear extremely important to support their development and the survival on the market too.

However, studies that address the relationship between small and medium enterprises (SMEs) and digital technologies, show a significant delay in the digitalization process and the potential seems not fully exploited yet. This gap can be traced back to the structural, but also cultural, characteristics of this type of companies.

Digital technologies include a very broad number of tools, which offer different opportunities and present different challenges for firms. These tools range from social media to the Industry 4.0 technologies, such as robotics, 3D printing, Internet of things, etc. Some of these technologies, such as social media, have been approached more diffusely by companies operating in business-to-consumer contexts and BtoB firms seem to lag behind in the digital transformation process.

In this regard, our objective is to provide some insights on the main drivers of the B2B SMEs digital transformation, starting from the main results of past studies produced by our research group. In particular, we deal with three topics: (1) the growing role of social media in the social selling and business relationships practices; (2) the adoption of a joint vision of the internationalization and digitalization trends, and finally (3) the e-learning as a marketing tool.

The methodological approach used has often provided for the integration of theoretical research with field research. In the last decade, in-depth corporate cases have been analyzed, meetings and direct interviews with managers and entrepreneurs were carried out. Furthermore, knowledge on these issues has also been gained thanks to the collaboration with entrepreneurial associations, which have allowed the creation of various empirical surveys, as well as the interaction with numerous companies that have begun to approach the "digital world" or in need of in-depth analysis on specific issues about the use of specific digital tools. In some cases, the members of the research group also had the opportunity to contribute to the development of strategic plans for businesses, aimed at improving web communication, and to the implementation of digital solutions and the performance analysis. This allowed the development of a relevant knowledge about the operational aspects that are faced by the SMEs.

The chapter is structured in two main parts. The first part highlights the main evolutionary aspects of digital transformation; moreover, the relationship between SMEs and digital technologies is analyzed. A focus is made on BtoB contexts. In the second part, the examination of three above-mentioned directions of the digital transformation is proposed: (a) online sales through social media; (b) digital internationalization, that is, the promotion and sale of foreign merchandise using Web tools; (c) the use of e-learning as a marketing tool.

The chapter concludes with some future perspectives of research.

2 The Digital Transformation

The ever-increasing impact that Information and Communication Technologies (ICTs) are having on organizations and society is contributing to what can be called a "digital revolution". The effects of these changes can be found in every area of human life, thus generating multiple and interrelated consequences. From a firm point of view, the implications of this revolution are pervasive and wide in scope, ranging from simple process innovations to business model innovations, with an increasingly convergence between products and services.[1] This is the reason why it seems quite difficult to give a definition of "digital transformation"; in general, we can refer to digital transformation as the changes in industrial and organizational processes and competencies, needed to grab the opportunities and face the challenges deriving from the new digital paradigm, which is enabled by different types of technologies, such as Internet of Things, Additive Manufacturing, Artificial Intelligence, etc. (Rindflesich et al. 2017). Most part of this transformation is tied to the phenomenon of "Industry 4.0", which, according to some authors, gave rise to the "fourth industrial revolution" (Schwab 2017).

The basis of these concepts is the application of digital technologies to the manufacturing sector in order to enable strategic, organizational, process and product innovations, with the aim to increase the competitiveness of both individual companies and economic system in general. The consulting firm Roland Berger defines the Industry 4.0 as "the set of technologies that will accompany the so-called fourth industrial revolution, based on the digitization and interconnection of all the production units present within an economic system". Two main components characterized this revolution: the digitalization and the connectivity (Bellagamba et al. 2018).

The path towards digitalization is not only linked to the understanding and adoption of technology by companies but it is the result of attitudes and managerial approaches focused on the ability to integrate these technologies in order to transform their business and their processes (Kane et al. 2015). It should, therefore, be stressed that it is the strategy, rather than technology that drives the digital transformation processes of companies (Pascucci and Temperini 2017).

---

[1] Analyzing the context of manufacturing companies, Rullani (2014) described a scenario in which "alongside the mass industry that continues to be such and becomes a low-cost global commodity industry, a new industry that instead seeks quality and therefore begins to offer the customer personalization, variety, meanings, experiences and guarantees that were once typical of services. Likewise, in the opposite direction but without passing through a material product—industrialized in production—and while guaranteeing a certain degree of flexibility—they provide users with standard services with virtually no production and transfer costs".

From this perspective, three macro trends are identified that are useful for understanding the "perimeter" of this evolution.

1. The first aspect is linked to the personalization of products and services made possible today thanks to the enabling technologies called the Internet of Things (IoT). These technologies are able to generate and make available in real time a large amount of data and information for example on customers, how to use and use of products-services, consumption and purchase behavior, but also biometric data, etc. These new forms of data-driven knowledge enable companies to design both new value propositions and new product-service solutions. IoT technologies can play an important role also in the post-purchase phases and in the management of customer relations; the high level of customization can indeed guarantee high customer satisfaction and customer loyalty performances.
2. Internet of Things and sensors applied to production machinery are giving life to the definition of new business models based on the offer of complex solutions, also the result of the reprogramming and reconfiguration of products that can be carried out remotely and in real-time.
3. Finally, the possibility of exchanging data and information between a large number of subjects is defining new business scenarios that will have wider opportunities for collaboration both inside and outside the production chain. The implications deriving from these assets are manifold: it is possible to identify new partners or to integrate more with existing partners, creating new sectors but also real business ecosystems (Iansiti and Euchner 2018).

As stated previously, digital transformation impacts on multiple dimensions of the enterprise. In the supply chain management, we can discuss three main points:

First. The automation of processes will be driven by 'Artificial Intelligence' and 'Big data' mutually supporting the requirement for sensor technologies as hardware on the one hand, as well as software on the other hand to collect, control, and process the enormous amount of available data.

Second. Interconnectivity is linked to internet technology and beyond the direct collaboration of supply chain partners by including real-time market developments, to include customer feedback from social media activities and share, monitor, and manage follow-up activities and decisions in real time. The change in supply chain design due to the influence of integrative technologies like (Big Data, IoT) and the opportunities for organisation to create synergies lead to the combination of 'normal' production and 'customer specific' production (additive manufacturing).

Third. Intelligent is related to simulations of supply chain events and supported by technologies. It is possible to create various scenarios in advance depending on future situations that result in a more efficient and effective supply chain control and the possibility to evaluate and eliminate risks before they occur (Bienhaus and Haddud 2017).

Marketing is one of the most affected areas: digital technologies are reshaping the marketing processes and the strategies, along with the following three main directions (Leeflang et al. 2014).

1. The ability to generate and strategically use insights on customers; digital technologies make the company able to draw on a large amount of data (often unstruc-
   tured) defined as *big data* (McKinsey et al. 2011) which, if properly collected,
   analsyzed and interpreted, provide the company with valuable knowledge on
   customer preferences and behaviors, along all the customer journey (Court et al.
   2009).
2. The growing importance of brand and firm reputation, as fundamental assets
   which require specific attention and skills, because of the explosion of user-
   generated contents in the context of social media and web 2.0 in general.
3. The measurement of digital marketing performance, with specific reference to
   the effects of this performance on the firm's economic-financial results. Although
   digital marketing tools have evolved very rapidly, this has not been the case in the
   web metrics, whose diffusion in the practice is very limited yet. The inability to
   measure results may cause a sort of "skepticism" in managers and entrepreneurs
   perceptions that contributes to further delaying the digitalization process (Bughin
   et al. 2008).

3   The SMEs in the Digital Scenario

In this changing context, the SMEs are among the realities whose digitalization
process presents the greatest complexities.

Currently, SMEs tend to concentrate their investments mainly on projects aimed
at digitizing basic processes, such as accounting and financial management, in order
to reduce costs and improve efficiency (Pascucci and Temperini 2017).

According to data provided by the European Commission (2017), more than 77%
of European SMEs with 10–249 employees have their own website, 57% of them
have on their websites more "advanced" features, such as a price list and content
customization. Still, the percentage of European SMEs from 10 to 249 present on at
least one social media is about 44%, while the companies that have practiced B2C
e-Commerce are around 7%.

In light of these data, there is a growing interest in the study of digital transformation processes in SMEs, also considering the economic relevance of SMEs
worldwide.

There are numerous contributions that identify the presence of a delay by small
businesses in the use of digital technologies (Morgan-Thomas 2015; Jones et al.
2014). This gap is generally due to a lack of financial resources and technical skills.

Moreover, since technological solutions are often designed and conceived having
in mind large organizations, it is not uncommon for SMEs to show little confidence
in the investments that go in this direction. It is therefore evident that the SMEs need
specific solutions and approaches, based on their structurasl, cultural and organizational peculiarities since the mere transfer of solutions designed for large companies
can rarely be effective.

Digital is still perceived as "one of the possible options" and not as the path to growth. The biggest problem that slows down the digital transformation process is the cultural one (Pascucci and Temperini 2017): on the one hand, the limited awareness of the opportunities that the Net and digital technologies can provide; on the other hand, the lack of managerial and operational skills that are indispensable for an effective implementation of those technologies. Hence the need to invest in training to develop the digital skills of the staff within the company or to introduce external professionals, who are able to manage the complexity and specificity of the issues related to digitization.

In this regard, external subjects—such as institutions, competence centers, research centers, and Digital Innovation Hubs—may have a strategic role as "activators" of SMEs digitalization processes, (Lee et al. 2010). The relationships among adopting firms and these external subjects result in the creation of "digital ecosystems". The term "digital ecosystem" can refer to different meanings because it has been used in a variety of different contexts (knowledge management, ICT, engineering, etc.). Brohman and Negi (2015) identify the following three different points of view in defining digital ecosystems:

1. The economic perspective, that defines digital ecosystems as a "useful metaphor for understanding the dynamics of corporate networks at regional and sectoral level and their interaction with and through information and communication technologies";
2. The technical view, according to which digital ecosystems are the digital counterpart of biological ecosystems; they are robust, scalable and self-organizing architectures that can solve complex and dynamic problems;
3. The ecological view, which defines it as "a digital environment populated by digital species or digital components that can be software components, applications, services, knowledge, business processes, and modules, training modules, contractual frameworks, laws, etc.".

As stated by Iansiti and Lakhani (2018):

> The point is that the economy has become a giant network. One of the great analogies for understanding this highly connected network of people and things is to think about it as a biological ecosystem [...] In economic as well as biologic systems, you want to have an ecosystem that is sustainable, something that works well in the long term, something that can scale.

4   B2B SMEs Digital Transformation's Trends

Despite the high volume of business generated by this type of companies, scholars' contributions to the study of the phenomenon of digitization in B2B still remain limited and a "B2B knowledge gap" exist (Lilien 2016). However, the B2B environment in recent years has been deeply marked "disruptive" in particular by the following factors (Grove et al. 2018):

1. The commoditization of quality. The technical and qualitative differences between competing offerings have been significantly reduced by the widespread adoption of total quality management methodologies, such as Six Sigma. As a result, the concept of quality assumes the meaning of commodity and in this regard, companies need to offer additional forms of value;
2. New technologies. In many industries, new technologies, such as cloud computing, mobile applications, and artificial intelligence, put certain business models at risk because they are able to offer less costly ways of achieving the same functionality;
3. The abundance of information related to the product. The current ease of access to information means that B2B customers are now able to conduct their own research before the formal sales process begins. Customers are less inclined to ask the supplier for general information on the characteristics of the product.

The majority of the studies focus on digital technologies and in particular on the new communication tools (such as social media) in the B2C marketing context. However, it is interesting to note that, while the delay of the B2B SMEs is evident (Michaelidou et al. 2011; Jussila et al. 2014), the spread of digital technologies by industrial companies is a growing trend. We have identified three major trends of B2B digital transformation, and they have represented three relevant research topics of the Authors in the last ten years.

4.1 Social Selling

In analyzing the relationship between social media and B2B, with specific reference to the practice of social selling, it should be noted that among the factors that most significantly affect the use of social media in B2B, those "contextual" elements clearly emerge. related to the organizational environment (Pascucci et al. 2018).

The use of social media is in fact driven mainly by the level of organizational knowledge on social media and by the type of management attitudes. Guesalaga (2016) shows that the greater the organizational competence and the commitment towards social media, the more the sales department is willing to use these platforms in sales activities. If managers are aware and active in using social media, they are more willing to support social media initiatives within the sales organization.

Some studies also note that an imitative effect linked to the use of social media can occur along the entire supply chain (Rapp et al. 2013).

In practice, organizations within a network exchange information and communicate frequently, which can lead companies to adopt imitative behavior. For example, if vendors develop strategies to promote the brand, which involve social media marketing initiatives, retailers may try to imitate these strategies in the supply chain in order to increase their success. In particular, Rapp et al. (2013) state that the use of social media reinforces relations between suppliers and resellers, especially in cases where the former are ambidextrous (that is, the search for quality of service and innovation for improvement) and when their brands they have a solid reputation. Furthermore, it is not just the comparison with customers and competitors that influences the use of social media by salespeople; the sense of belonging to specific virtual communities could play an important role (Wang et al. 2016).

In the literature, the phenomenon of B2B social selling has been addressed in particular by two perspectives: social selling at individual salesperson level (Moore et al. 2015; Schuldt and Totten 2015) and social selling at the organizational level (Agnihotri et al. 2012; Andzulis et al. 2012).

4.1.1 Social Selling at the Individual Person Level

According to Moore et al. (2015), B2B salespeople also use social media tools to a greater extent than sales personnel in the consumer sector.

In particular, some social platforms, such as professional networking tools and instant messaging applications, are more common among B2B salespeople (Moore et al. 2015). For example, studies show that B2B vendors tend to use blogs, professional networking sites (e.g., LinkedIn), interactive broadcasting, webinar tools, and presentation sharing sites (e.g. Slideshare) to a greater extent than their B2C colleagues.

With another approach, more aimed at understanding the degree of use of social media in sales practices; Rodriguez et al. (2016) for example measures the increase in the use of social media over time with the aim of identifying the presence of business opportunities and key decision makers.

Other studies investigate the usefulness of social media in carrying out sales work and its integration in daily work (Agnihotri et al. 2016, 2017; Itani et al. 2017). Rapp et al. (2013) measure the use of social media, including social networking activities and behaviors, such as monitoring competitors, providing information to customers, monitoring events and development in the sector.

In this way, this group of studies focuses primarily on the overall use of social media technologies in sales and provides limited information on how salespeople actually take advantage of social media in their work. Despite a growing tendency to measure the general degree of use of social media in sales (see Agnihotri et al. 2016, 2017; Itani et al. 2017), most studies still use different constructs, which means that measures with a strong theoretical basis are largely missing.

With a final analysis perspective of this dimension, the researchers provided some information on how salespeople use social media in sales practices and their social selling activities (Bocconcelli et al. 2017; Lacoste 2016; Rollins et al. 2014; Wang et al. 2016). Salespeople seem to use social media, in particular, to gather information and better understand customers (Lacoste 2016). Salespeople may use social media to search for the right contact and get various types of information about potential customers, such as the characteristics of their network, the degree of experience and interests, which can be useful for preparing the start of the relationship. Furthermore, social media allow salespeople to actively pursue networking opportunities with targeted customers and relevant customer stakeholders (Bocconcelli et al. 2017; Lacoste 2016).

The development of a solid network through social media allows sellers to find potential customers through their existing connections, thus exploiting the credibility of the network. In fact, social media offers the possibility of collecting information even beyond the seller-customer relationship, exploiting both the internal (within the customer's organization) and external networking (i.e. with suppliers, competitors, etc.) (Lacoste 2016). In this regard, Bocconcelli et al. (2017) show that social media make it possible to establish the first contact with potential business partners, provide relevant information in a less formal way (e.g. YouTube videos), also by using word of mouth.

4.1.2   Social Selling at the Organizational Level

A significant amount of studies in this area of research focuses on organizational aspects related to social selling. In particular, this category includes all the jobs that specifically concern the implementation at the company level of sales that are enabled by social media.

The analysis of these studies reveals two important organizational aspects related to social selling, namely the social selling strategy and organizational activities consisting of tools, activities, and processes aimed at supporting the sales force of the sellers.

The literature has amply underlined the importance of developing and communicating a clear social selling strategy, i.e. a company-wide policy that provides indications on how to operate in the field of social selling (Agnihotri et al. 2012; Andzulis et al. 2012). In particular, the definition of a social selling strategy goes beyond establishing a mere presence on social media or implementing sales tactics conceived in a disjoint way. Furthermore, social media seem to influence the organization's sales capacity to create opportunities and manage relationships, so companies should think carefully about its systematic integration with more *traditional* consolidated sales processes. In fact, social media seem to have the potential to influence every traditional step of the sales process that goes from customer understanding, to client approach, and to discovery, presentation, closure, and follow-up (Andzulis et al. 2012; Marshall et al. 2012; Moncrief et al. 2015).

4.2   Digital Internationalization

Internationalization and digitalization represent two of the most studied and debated themes among researchers, practitioners and also policymakers of the last decades.

Their importance for the competitiveness of companies and of entire economic systems is such that each of them has generated numerous publications. Only recently, however, the two topics have begun to be treated jointly, in order to understand what relationship there is between the two phenomena (Pascucci and Temperini 2017).

Studies in this regard can be divided into the following two categories.

The first one considers internationalization as an influential factor in digitalization. The effects of internationalization on the adoption of digital technologies can be explained by recourse to the theory of resource-based view, in the sense that companies operating in foreign markets need some types of resources and skills that technologies can help to develop so that internationalized companies would have a greater incentive to adopt such technologies. However, the analysis of these studies shows non-univocal results (Bayo-Mariones and Lera-Lòpez 2007; Zhu et al. 2003; Zhu and Kraemer 2005).

The second one considers digitalization as an influential factor on the firms' international performance, focusing on the benefits that Internet adoption can provide for the development of exports and penetration into foreign markets. This is the most nurtured research line, in which the studies agree on the positive influence of digitalization on the firms' export performance: the Internet is seen as a facilitator, or as an enabler of internationalization or as a driver of new international opportunities (Bell and Loane 2010).

The Web lowers the costs of marketing and promotional research, and allows greater visibility also in geographically distant markets, without having to be physically present (Karakaya and Karakaya 1998; Tiessen et al. 2001).

Several authors have placed particular attention on SMEs and the benefits they can derive from using the Internet (Hamill 1999; Sinkovics et al. 2013). The Web allows them to coordinate and maintain communication between different geographical areas in a simple and economical way, to develop and maintain relationships with customers, business partners, and foreign suppliers, as well as reducing barriers to internationalization (Hamill and Gregory 1997).

Considering the fact that the main advantages related to the use of digital technologies are the access and management of information, as well as communication, the contribution they can provide to reducing the barriers to exports is evident.

The use of the online channel for the collection of information and the use of innovative software for the processing and management of such information also allows smaller companies to acquire a deep knowledge of foreign markets, from a macro and a micro point of view, at relatively low costs, thus reducing the perceived risk and uncertainty typical of operating in unknown and distant markets (Mathews et al. 2016). Finally, the idea that digitalization and the advent of the Internet can favor the internationalization processes of B2B SMEs, by eliminating the physical distance between seller and customer, is now widely shared.

As previously stated, recent attention has shifted to the opportunities that, in this context, can provide social media (Marinelli 2017), as a direct channel of communication with the foreign customer (Broncanello and Tremiterra 2015).

With this in mind, it is important to underline that although the Internet has allowed the "geographical distance" and "temporal distance" to be reduced, shortening the distribution channels and favoring direct communication with the user from any part of the world, digital tools have not completely canceled that distance of a cultural nature. The cultural differences between countries are also maintained on the Net, influencing the preferences, behaviors, and mechanisms of interaction with foreign customers (Gregori et al. 2016).

4.3   The Use of E-learning for Marketing Purposes

The growing interest that has long been encountered with regard to e-learning is motivated by the important advantages that this formative method allows to obtain. In fact, through the use of ICT, it is possible to considerably extend learning opportunities and increase the usability of *knowledge* responding to a widespread need for continuous learning that is felt by individuals and organizations.

With specific reference to companies, it is believed that e-learning can play a significant role for their development, even if different diffusion problems can be observed (Gregori and Temperini 2007, 2009). The high flexibility that distinguishes it can, first of all, allow access to training for companies, even those of a smaller size, allowing them to increase their internal skills and affect their competitiveness. Furthermore, e-learning can contribute to enhancing and increasing the organizational cognitive heritage, stimulating the creation and diffusion of new knowledge, starting from the existing ones; in this sense, the integration between e-learning and knowledge management systems (Wild et al. 2002; Capucci 2005; Maier and Schmidt 2007) is found in several cases.

In the scientific literature on management issues, e-learning has been considered mainly as a tool for human resources management and knowledge management, and has also been observed as a means of internal communication (Iacono 2001). On the other hand, little attention was paid to the use of this tool from a marketing point of view, although the potential for use appears particularly interesting, in light of the development and diffusion of digital technologies.

The contribution of our research team on the subject in question is characterized by an original approach, placing the main objective of observing the possible role of e-learning in business communication strategies, and also the contribution it can offer in initiatives branding aimed at developing relationships in the market.

To this end, the case of iGuzzini, a medium enterprise which produces lighting tools and is based in Marche region (Italy), has been analyzed. The company offer includes several product lines divided into the following main areas: (a) indoor lighting systems; (b) outdoor lighting systems; (c) light management systems; (d) special products (for example, direct light system for urban areas or professional fluorescent systems and suspensions). The solutions are aimed at lighting for urban furniture, the tertiary sector, museums, commercial spaces, and reception facilities. Its products are sold internationally through a sales network (also composed of exclusive distributors and some branches) that extends to over 60 countries.

iGuzzini has characterized its presence on the Web by implementing a technological platform (which is named Lightcampus) dedicated to offering online training services on topics related to the quality of light and lighting technology.

The examination of this case study allows observing the possible role that e-learning can play for communication and marketing purposes. In particular, the development of the e-learning platform aimed at satisfying a widespread need and desire for learning with regard to lighting issues allows the company to:

- Communicate the products with an innovative approach that favors high levels of interaction and informative detail;
- Promote brand awareness and strengthen the link to it, also making use of the creation of a virtual community;
- Get in touch and interact with important players in the market, in this case especially with the "influencers" or "prescribers" (architects and lighting engineers), with the possibility of obtaining important data and information on that players, useful for the implementation of policies of relational marketing based on the integration of e-learning and CRM systems.

It is evident that the observed model can be a valid reference for companies operating in other sectors, even for those that address the consumer markets. The challenge for companies lies in identifying the contents and e-learning methods that best meet the needs of users who correspond to the target to be achieved.

The prospect of the use of e-learning in terms of communication and marketing is particularly interesting in light of the potential that can be expressed through integration and convergence with other IT and Web technologies.

5 Some Future Research Directions

Digital transformation has attracted growing attention of both scholars and practitioners, given its enormous potential impact on products, services, processes and business models. Despite this fact, several gaps exist and further theoretical and empirical research is needed, in particular with regard to BtoB SMEs.

Digital technologies create a *potential* that firms have to transform in *reality* but this requires resources, skills, competencies that are often lacking especially in the context of smaller firms.

One of the most critical areas is digital analytics for two main reasons. First of all, proving the added value that digital tools bring to companies, showing the tangible impact on firm's performance, is a fundamental aspect in order to stimulate the adoption and the use of technological solutions and promote a better *digital culture* even in smaller companies. This appears even more relevant in the case of BtoB contexts. Second, digital tools can help to improve the competitiveness of SMEs increasing the data and the information needed to make efficient and effective decisions; data has been called *the oil* of the digital economy (Wedel and Kannan 2016). Digital transformation is a vehicle for new technologies which provide companies with the opportunity to generate, acquire, aggregate, analyze and monitor a huge volume of real-time data that can come from multiple sources. When adopted with a strategic approach, these data can be transformed into insights that allow companies to make decisions in an increasingly complex scenario and optimize marketing spending. However, these data have to be transformed in actionable knowledge by firms and this requires completely new skills. According to a Gartner's study (Levy 2015), the difficulty to finding talent with these skills is the main barrier towards implementing marketing analytics and this also present a challenge for educators and for academia.

Moreover, the mere adoption of digital technologies by itself is not a sufficient condition for the improvement of company performance, since the effects depend on several factors: the type of technology considered, the observation period, the presence of complementary resources in the company, such as the organizational changes and human resource skills. According to recent research conducted by MIT Sloan Management Review in collaboration with Deloitte on more than 4800 managers of companies in different sectors and Countries, it has become clear that the advantages of digital technologies do not lie in the technologies themselves, but in the ways in which companies integrate these technologies to transform their business and processes (Kane et al. 2015). The conclusion reached by scholars is as follows: it is the strategy rather than technology that drives the digital transformation of companies. In fact, what distinguishes the most mature companies from the digital point of view is the implementation of a clear digital strategy, combined with an organizational culture and an entrepreneurial approach open to collaboration, risk, and experimentation (Quinton et al. 2017, Li et al. 2018).

Most parts of the skills and complementary resources needed in order to succeed in the digital transformation process are intangible: the role of "intangibles" within the digital transformation processes is another fruitful future research direction. It is argued that companies, in order to start successful digitization projects, must necessarily invest above all, in "soft factors", such as know-how, skills, relational capital, and experience (Caroli and Van Reene 2001). In particular, the role of external relationships seems to be very important for SMEs and for BtoB contexts. In the digital age BtoB relationships are changing and evolving in new forms but little is known on the nature and consequences of these transformations; how these technologies are changing the relationships a company has with its customers, its suppliers or with other actors remain still unclear (Pagani and Pardo 2017). In particular, the implications of digitalization on the "dark side" of BtoB relationships (Abosang et al. 2016) need to be studied because this area has received scarce attention.

It is worth nothing that managers need to work toward the development of an open-innovation capability, which comprises four value processes: value provision, value negotiation, value realization, and value partake (Appleyard and Chesbrough 2017; Chesbrough et al. 2018).

Firms can develop processes to seek out and transfer external knowledge into their own innovation activities. They can also create channels to move unutilized internal knowledge to other organizations in the surrounding environment.

References

Abosang, I., Yen, D. A., & Barnes, B. R. (2016). What is the dark about the dark side of business relationships? *Industrial Marketing Management, 55,* 5–9.

Agnihotri, R., Dingus, R., Hu, M. Y., & Krush, M. T. (2016). Social media: influencing customer satisfaction in B2B sales. *Industrial Marketing Management, 53,* 172–180.

Agnihotri, R., Kothandaraman, P., Kashyap, R., & Singh, R. (2012). Bringing "social" into sales: the impact of salespeople's social media use on service behaviors and value creation. *Journal of Personal Selling & Sales Management, 32*(3), 333–348.

Agnihotri, R., Trainor, K. J., Itani, O. S., & Rodriguez, M. (2017). Examining the role of sales-based CRM technology and social media use on post-sale service behaviors in India.*Industrial Marketing Management, 81,* 144–154.

Andzulis, J. M., Panagopoulos, N. G., & Rapp A. (2012). A review of social media and implications for the sales process. *Journal of Personal Selling & Sales Management, 32*(3), 305–316.

Appleyard, M., & Chesbrough, H. (2017). The dynamics of open strategy: from adoption to reversion. *Long Range Planning, 50*(3), 310–321.

Bayo-Mariones, A., & Lera-Lòpez, F. (2007). A firm-level analysis of determinants of ICT adoption in Spain. *Technovation, 27,* 352–366.

Bell, J., & Loane, S. (2010). New-wave global firms: Web 2.0 and SME internationalization. *Journal of Marketing Management, 26*(3-4), 213–229.

Bellagamba, A., Gregori, G. L., Pascucci, F., Perna, A., & Sabatini, A. (2018). Industria 4.0: Non solo una rivoluzione tecnologica. In A.A.V.V (Eds.), *Le competenze per costruire il futuro.* Roma: Edizioni di Comunità.

Bienhaus, F., & Haddud, A. (2017). Procurement 4.0: factors influencing the digitisation of procurement and supply chains. *Business Process Management, 24*(4), 965–984.

Bocconcelli, R., Cioppi, M., & Pagano, A. (2017). Social media as a resource in SMEs' sales process. *Journal of Business & Industrial Marketing, 32*(5), 693–709.

Brohman, K., & Negi, B. (2015). Co-creation of value in digital ecosystems: A conceptual framework. In *AMCIS.*

Broncanello, I., & Tremiterra, M. (2015). Il ruolo delle risorse e delle competenze organizzative dell'impresa nell'utilizzo e implementazione del web 2.0: Il caso delle imprese della strada dell'olio in Umbria. *Mercati e Competitività, 4,* 105–131.

Bughin, J., Shenkan, A. J., & Singer, M. (2008). How poor metrics undermine digital marketing. *The McKinsey Quarterly, 10,* 1–5.

Capucci, U. (2005). E-learning, un'importante supporto del knowledge management. *FOR (Rivista per la formazione), 63.*

Caroli, E., & Van Reenen, J. (2001). Skill-biased organizational change? Evidence from a panel of British and French establishment. *The Quarterly Journal of Economics, 116*(4), 1449–1492.

Chesbrough, H., Lettl, C., & Ritter, T. (2018). Value creation and value capture in open innovation. *Journal of Product Innovation Management, 35*(6), 930–938.

Court, D., Elzinga, D., Mulder, S., & Vetvik, O. J. (2009). The consumer decision journey. *McKinsey Quarterly, 3,* 1–11.

European Commission. (2017). Integration of digital technology. In *Europe's digital progress report 2017.*

Gregori, G. L., Pascucci, F., & Cardinali, S. (2016). Internazionalizzazione digitale. In *Come vendere online nei mercati esteri.* Milano: Franco Angeli.

Gregori, G. L., & Temperini, V. (2007). Problematiche di sviluppo dell'e-learning in Italia: I risultati di un'indagine empirica. *Economia e diritto del terziario, 1.*

Gregori, G. L., & Temperini, V. (2009) E-learning e formazione per le PMI calzaturiere: quali nuove prospettive. *Piccola Impresa/Small Business, 1.*

Grove, H., Sellers, K., Ettenson, R., & Knowles, J. (2018). Selling solutions isn't enough. *MIT Sloan Management Review*, Fall 55–59.

Guesalaga, R. (2016). The use of social media in sales: individual and organizational antecedents, and the role of customer engagement in social media. *Industrial Marketing Management, 54,* 71–79.

Hamill, J. (1999). Internet editorial: export guides on the net. *International Marketing Review, 15*(5), 434–436.

Hamill, J., & Gregory, K. (1997). Internet marketing in the internationalization of UK SMEs. *Journal of Marketing Management, 13*(1–3), 9–28.

Iacono, G. (2001). *Dal knowledge management alla e-enterprise.* Milano: Franco Angeli.

Iansiti, M., & Euchner, J. (2018). Competing in ecosystems. *Research-Technology Management, 61*(2), 10–16.

Iansiti, M., & Lakhani, K. R. (2018). Managing our hub economy. *Harvard Business Review, 96*(1), 17.

Itani, O. S., Agnihotri, R., & Dingus, R. (2017). Social media use in B2B sales and its impact on competitive intelligence collection and adaptive selling: examining the role of learning orientation as an enabler. *Industrial Marketing Management, 66,* 64–79.

Jones, P., Packham, G., Beynon-Davies, P., Simmons G., & Pickernell D. (2014). An Exploration of the attitudes and strategic responses of sole-proprietor micro-enterprises in adopting ICT. *International Small Business Journal, 32*(3), 285-306

Jussila, J. J., Kärkkäinen, H., & Aramo-Immonen, H. (2014). Social media utilization in business-to-business relationships of technology industry firms. *Computers in Human Behaviour, 30,* 606–613.

Kane, G. C., Palmer, D., Phillips, N., & Kiron, D. (2015). Is your business ready for a digital future? *MIT Sloan Management Review, 56*(4), 37–44.

Karakaya, F., & Karakaya, F. (1998). Doing business on the Internet. *SAM Advanced Management Journal,* Spring, 10–14.

Lacoste, S. (2016). Perspectives on social media and its use by key account managers. *Industrial Marketing Management, 54,* 33–43.

Lee, S., Park, G., Yoon, B., & Park, J. (2010). Open innovation in SMEs—An intermediated network model. *Research Policy, 39,* 290–300.

Leeflang, P. S. H., Verhoef, P. C., Dahlstrom, P., & Freundt, T. (2014). Challenges and solutions for marketing in a digital era. *European Management Journal, 32,* 1–12.

Levy, H. P. (2015). Marketers embrace analytics and look for talent. Available at www.gartner.com.

Li, L., Su, F., Zhang, W., & Mao, J. Y. (2018). Digital transformation by SME entrepreneurs: A capability perspective. *Information System Journal, 28,* 1129–1157.

Lilien, G. L. (2016). The B2B knowledge gap. *International Journal of Research in Marketing, 33*(3), 543–556.

Maier, R., & Schmidt, A. (2007). Characterizing knowledge maturing: A conceptual process model for integrating E-learning and knowledge management. In *4th Conference Professional Knowledge Management* (WM 07). Potsdam, Germany.

Marinelli, L. (2017) Una visione globale dei social media. In Gregori, G. L., Pascucci, F., & Cardinali, S. (Eds.), *Internazionalizzazione digitale. Come vendere online sui mercati esteri.* Milano: Franco Angeli.

Marshall, G W., Moncrief, W. C., Rudd, J. M., & Lee, N. (2012). Revolution in sales: The impact of social media and related technology on the selling environment. *Journal of Personal Selling & Sales Management, 32*(3), 349–363.

Mathews, S., Bianchi, C., Perks, K. J., Healy, M., & Wickramasekera, R. (2016). Internet marketing capabilities and international market growth. *International Business Review, 25,* 820–830.

McKinsey, G., et al. (2011). *Big data: The next frontier for innovation, competition and productivity.* McKinsey Global Institute.

Michaelidou, N., Siamagka, N. T., & Christodoulides, G. (2011). Usage, barriers and measurement of social media marketing: An exploratory investigation of small and medium B2B brands. *Industrial Marketing Managemen, 40,* 1153–1159.

Moncrief, W. C., Marshall, G. W., & Rudd, J. M. (2015). Social media and related technology: Drivers of change in managing the contemporary sales force. *Business Horizons, 58*(1), 45–55.

Moore, J. N., Raymond, M. A., & Hopkins, C. D. (2015). Social selling: A comparison of social media usage across process stage, markets, and sales job functions. *Journal of Marketing Theory and Practice, 23*(1), 1–20.

Morgan-Thomas, A. (2015). Rethinking technology in the SME context: Affordances, practices and ICTs. International Small Business Journal: Researching Entrepreneurship, 34(8), 1122-1136

Pagani, M., & Pardo, C. (2017). The impact of digital technology on relationships in a business network. *Industrial Marketing Management, 67,* 185–192.

Pascucci, F., & Temperini, V. (2017). Trasformazione digitale e sviluppo delle PMI. Giappichelli.

Pascucci, F., Ancillai, C., & Cardinali, S. (2018). Exploring antecedents of social media usage in B2B: A systematic review. *Management Research Review, 41*(6), 629–656.

Quinton, S., Canhoto, A., Molinillo, S., Pera, R., & Budhathoki, T. (2017). Conceptualising a digital orientation: Antecedents of supporting SME performance in the digital economy. *Journal of Strategic Marketing*.

Rapp, A., Beitelspacher, L. S., Grewal, D., & Hughes, D. E. (2013). Understanding social media effects across seller, retailer, and consumer interactions. *Journal of the Academy of Marketing Science, 41*(5), 547–566.

Rindflesich, A., O'Hern, M., & Sachdev, V. (2017). The digital revolution, 3D printing, and innovation as data. *Journal of Product Innovation Management, 34*(5), 681–690.

Rodriguez, M., Ajjan, H., & Peterson, R. M. (2016). Social media in large sales forces: An empirical study of the impact of sales process capability and relationship performance. *Journal of Marketing Theory and Practice, 24*(3), 365–379.

Rollins, M., Nickell, D., & Wei, J. (2014). Understanding salespeople's learning experiences through blogging: A social learning approach. *Industrial Marketing Management, 43*(6), 1063–1069.

Rullani, E. (2014). Manifattura in transizione. In *Sinergie, 93* (pp. 141–152). January–April, 2014.

Schuldt, B. A., & Totten, J. W. (2015). Application of social media types in the sales process. *Academy of Marketing Studies Journal, 19*(3), R230.

Schwab, K. (2017). *The fourth industrial revolution*. New York: Crown Business.

Sinkovics, N., Sinkovics, R. R., & Jean, R. (2013). The Internet has an alternative path to internationalization? *International Marketing Review, 30*(2), 130–155.

Tiessen, J. H., Wright, R. W., & Turner, I. (2001). A model of e-commerce use by internationalizing SMEs. *Journal of International Management, 7,* 211–233.

Wang, Y., Hsiao, S. H., Yang, Z., & Hajli, N. (2016). The impact of sellers' social influence on the co-creation of innovation with customers and brand awareness in online communities. *Industrial Marketing Management, 54,* 56–70.

Wedel, M., & Kannan, P. K. (2016). Marketing analytics for data-rich environments. *Journal of Marketing, 80,* 97–121.

Wild, R. H., Griggs, K. A., & Downing T. (2002). A framework for e-learning as a tool for knowledge management. *Industrial Management & Data Systems* (Vol. 102).

Zhu, K., & Kreamer, K. L. (2005). Post-adoption variations in usage and value of e-business by organizations: cross-country evidence from retail industry. *Information System Research, 16*(1), 61–84.

Zhu, K., Kreamer, K. L., & Xu, S. (2003). Electronic business adoption by European firms: a cross-country assessment of the facilitators and inhibitors. *European Journal of Information Systems, 12,* 251–268.

Printed in the United States
By Bookmasters